Medical Imaging

Artificial Intelligence, Image Recognition, and Machine Learning Techniques

Medical Imaging

Artificial Intelligence, Image Recognition, and Machine Learning Techniques

K.C. Santosh, Sameer Antani,
D.S. Guru, and Nilanjan Dey

CRC Press is an imprint of the
Taylor & Francis Group, an informa business

CRC Press
Taylor & Francis Group
52 Vanderbilt Avenue,
New York, NY 10017

Printed on acid-free paper

International Standard Book Number-13: 978-0-367-13961-2 (Hardback)

Library of Congress Cataloging-in-Publication Data

Names: Santosh, K. C., editor.
Title: Medical imaging : artificial intelligence, image recognition, and machine learning techniques / [edited by] KC Santosh, Sameer Antani, DS Guru, Nilanjan Dey.
Description: Boca Raton : Taylor & Francis, a CRC title, part of the Taylor & Francis imprint, a member of the Taylor & Francis Group, the academic division of T&F Informa, plc, 2020. | Includes bibliographical references and index.
Identifiers: LCCN 2019015642| ISBN 9780367139612 (hardback : acid-free paper) | ISBN 9780429029417 (e-book)
Subjects: LCSH: Diagnostic imaging.
Classification: LCC RC78.7.D53 M43215 2020 | DDC 616.07/54--dc23
LC record available at https://lccn.loc.gov/2019015642

Visit the Taylor & Francis Web site at
http://www.taylorandfrancis.com

and the CRC Press Web site at
http://www.crcpress.com

Contents

Preface

This book aims to provide advanced or up-to-date techniques in medical imaging through the use of artificial intelligence (AI), image recognition (IR), and machine learning (ML) algorithms/techniques. An image or a picture is worth a thousand words; which means that image recognition can play a vital role in medical imaging and diagnostics, for instance. The data/information in the form of image, i.e. a set of pixels, can be learned via AI, IR, and ML, since it is impossible to employ experts for big data. The book covers several different topics, such as tuberculosis (TB) detection; radiologic and urologic applications; epileptic seizures detection; histology classification of non-small cell lung cancer; osteoarthritis classification (using knee joint X-ray); non-proliferative diabetic retinopathy lesions classification; fractured bone detection and labeling (using CT images); usefulness of 3D imaging (a quick review); and pathological medical imaging and segmentation.

In Chapter 1, authors discuss the stacked generalization of models for TB detection in chest radiographs. TB is an airborne infection and a common cause of death related to antimicrobial resistance. In resource-constrained settings, where there is a significant lack of expertise in interpreting radiology images, there is a need of image-analysis-based computer-aided diagnosis (CADx) tools. Such tools have gained significance because they offer a promise to alleviate the human burden in screening in countries that lack adequate radiology resources. Very specifically, authors reported the use of convolutional neural networks (CNN), a class of deep learning (DL) models. We observed that such tools deliver promising results on visual recognition tasks with end-to-end feature extraction and classification. Besides, ensemble learning (EL) methods combine multiple models to offer promising predictions because they allow the blending of intelligence from different learning algorithms.

In Chapter 2, the authors provide a thorough idea on how artificial intelligence (AI) tools can help in medical imaging, using radiologic and urologic applications. The authors are convinced of the fact that diagnostic errors account for approximately 10% of patient deaths, and between 6 and 17% of adverse events occurring during hospitalization. At a total of ~20 million radiology errors per year, and 30,000 practicing radiologists, this averages to just under 700 errors per practicing radiologist. Errors in diagnosis have been associated with clinical reasoning, including: intelligence, knowledge, age, affect, experience, physical state (fatigue), and gender (male predilection for risk taking). These factors, and the limited access to radiologic specialists for up to 2/3 of the world, encourage a more urgent role for the use of AI in medical imaging, a huge focus of which is machine learning. Also, operator dependency in radiologic procedures, particularly sonography, has led researchers to develop automated image interpretation techniques similar to those used for histopathology. AI now allows for the connection of image analysis with diagnostic outcome, in real time. AI has the potential to assist with care, teaching, and diagnosis of illness. According to the market research firm Tractica, the market for virtual digital assistants worldwide will reach $16 billion by 2021. The term "machine learning" as it applies to *radiomics* is used to describe high throughput extraction of quantitative imaging features with the intent of creating minable databases from radiological images.

In Chapter 3, the authors discuss the early detection of epileptic seizures, which is based on scalp electroencephalography (EEG) signals. Their research aims to realize a seizure detector using the empirical mode decomposition (EMD) algorithm and a machine learning–based classifier that is robust enough for practical applications. They have conducted exhaustive tests on EEG data of 24 pediatric patients who suffered from intractable seizures. Their tool may serve as a potential avenue for real-time seizure detection.

In Chapter 4, authors reported the usefulness of fractals in histology classification of non-small cell lung cancer (NSCLC). This type of cancer accounts for 85% of all the lung cancers. Noninvasive identification of the histology of NSCLC aids in determining the appropriate treatment approaches. In this study, the authors observed the usage of radiomics with application of fractals for decision-making: histology classification of NSCLC using lung CT images. Again, their study suggests that fractals can play a vital role in radiomics, providing information about the gross tumor volume (GTV) structure, and also helping characterizing the tumor.

In Chapter 5, the authors explain the use of multiple features to classify osteoarthritis (OA) in knee joint X-ray images. OA is a commonly occurring disease in the joints of the knee, hip, and hands. It results in a loss of cartilage. Affected patients will experience severe pain, stiffness, and a grating sensation during the movement of the joints. The authors reported that the use of several different features, such as edge, curvature, and textures, could improve the performance in classification (normal and/or abnormal), where conventional machine learning classifiers are used.

In Chapter 6, the authors explained how non-proliferative diabetic retinopathy (NPDR) lesions could be detected and classified. Diabetic retinopathy began with a leakage of blood or fluid from the retinal blood vessels and it damages the retina. NPDR is an early stage of diabetic retinopathy and it is categorized by three stages: mild, moderate, and severe. These were tested, and the authors achieved a classification accuracy of 94% using artificial neural network.

In Chapter 7, the authors explain the use of image segmentation so that image region labeling becomes easier. In their study, they discuss bone fracture detection and labeling in computed tomography (CT) images. CT images are a crucial resource for assessing the severity and prognosis of bone injuries caused by trauma or accident. Similarly, fracture detection is a very challenging task. In their work, the authors developed a computer-aided diagnosis (CAD) system, which not only precisely extracts and assigns unique labels to each fractured piece by considering patient-specific bone anatomy, but also effectively removes unwanted artifacts (like flesh) surrounded by bone tissues. In their tests (real patient-specific CT images), they have reported the maximum possible accuracy of 95%.

In Chapter 8, the authors provide a systematic review on 3D imaging in biomedical applications. The volume visualization or 3D imaging field is vibrant and one of the fastest growing fields in scientific visualization. It is focused on creating high-quality 3D images from acquired volumetric datasets to gain insights into underlying data. In this work, the authors primarily review detailed information about the state-of-the-art volume visualization techniques majorly applied in the biomedical field. Besides, they provide commonly used tools and libraries that are employed for volume visualization. Further, several applications are discussed.

In Chapter 9, the authors discuss the evolution of the digital sliding of pathology in medical imaging. In general, they point out the evolution in the digitalization of pathological slides and explain the advantages of pathology practices in the prediction of diseases, in minimizing efforts, and in clarifying disease information via diagnosis. For example, examining tiny tissue uncovers data that could empower the pathologist to render an accurate analysis and provide help with treatments.

In Chapter 10, the authors provide a quick review on pathological medical segmentation, which is based on parametric techniques. In their study, several different segmentation techniques are considered. Further, authors point out the comparison among the techniques (publicly available) and help readers find an appropriate one.

MATLAB® is a registered trademark of The MathWorks, Inc. For product information, please contact:

The MathWorks, Inc.
3 Apple Hill Drive
Natick, MA 01760-2098 USA
Tel: 508 647 7000
Fax: 508-647-7001
E-mail: info@mathworks.com
Web: www.mathworks.com

Editors

K.C. Santosh (Senior member, IEEE) is an assistant professor and graduate program director for the Department of Computer Science, University of South Dakota (USD). Also, Dr. Santosh is an associate professor (visiting) for the School of Computing and IT, Taylor's University Before joining the USD, Dr. Santosh worked as a research fellow at the U.S. National Library of Medicine (NLM), National Institutes of Health (NIH). He worked as a post-doctoral research scientist at the LORIA research centre, Université de Lorraine in direct collaboration with ITESOFT, France (industrial partner). He received his PhD diploma in computer science from INRIA – Université de Lorraine (France), and his MS in computer science from Thammasat University (Thailand). Dr. Santosh has demonstrated expertise in pattern recognition, image processing, computer vision, artificial intelligence, and machine learning with various applications in medical image analysis, graphics recognition, document information content exploitation, and biometrics. He has published more than 120 peer-reviewed research articles and two authored books (Springer); he has also edited several books (Springer, Elsevier, and CRC Press), journal issues (Springer), and conference proceedings (Springer). Dr. Santosh serves as an associate editor of the *International Journal of Machine Learning and Cybernetics* (Springer). For more information, please visit: http://kc-santosh.org.

Sameer Antani is a staff scientist and (acting) chief of the communications engineering branch and the computer science branch, respectively, at the Lister Hill National Center for Biomedical Communications, an intramural R&D division of the National Library of Medicine (NLM) part of the National Institutes of Health (NIH) in Bethesda, Maryland. He is a versatile senior researcher, leading several scientific and technical research explorations to advancing the role of computational sciences and engineering in biomedical research, education, and clinical care. His research applies and studies methods for explaining the behavior of machine intelligence methods in automated decision support in biomedical applications. For this, he draws on his expertise in biomedical image informatics, automatic medical image interpretation, information retrieval, computer vision, and related topics in computer science and engineering technology. His contributions include automated screening for high burden diseases such as (i) Tuberculosis (TB) in HIV positive patients using digital chest X-ray image analysis; (ii) cervical cancer in women using analysis of acetowhitened images of the cervix, and whole slide images of liquid-based Pap smears and histopathology images from cervical biopsies; and (iii) automated methods for detecting malaria parasites in microscopic images of thick and thin blood smears. Other contributions include functional MRI (fMRI) simulation for brain research and similarity retrieval; analysis of ophthalmic fundus images for glaucoma, and the OPEN-i® biomedical image retrieval system, which provides text and visual search capability to retrieve over 3.7 million images from about 1.2 million of NLM's PubMed Central® Open-Access articles and other image datasets. Dr. Antani is a senior member of the International Society of Photonics and Optics (SPIE), the Institute of Electrical and Electronics Engineers (IEEE). He serves as the vice chair for computational medicine on the IEEE Computer Society's Technical Committee on Computational Life Sciences (TCCLS), and on the editorial boards of *Heliyon* and *Data*.

Dr. Antani is a senior member of the International Society of Photonics and Optics (SPIE), the Institute of Electrical and Electronics Engineers (IEEE), and the IEEE Computer Society. He serves as the vice chair for computational medicine on the IEEE Technical Committee on Computational Life Sciences (TCCLS), and as an associate editor for the *IEEE Journal of Biomedical and Health Informatics*

D.S. Guru is a professor at the Department of studies in computer science. He is known for his contributions to the fields of image processing and pattern recognition. He is a recipient of the BOYSCAST Fellowship awarded by the Department of science and technology, Govt of India, and also of the Award for Research Publications by the Department of science and technology, Karnataka Government. He has been recognized as the best ethical teacher in higher learning by Rotary North Mysore. He has supervised more than 15 PhD students and is currently supervising many more. He holds rank positions in both bachelors and masters education. He earned his doctorate from the University of Mysore and did his post-doctoral work at PRIP Lab, Michigan State University. He has been a reviewer for international journals by Elsevier Science, Springer, and IEEE Transactions. He has chaired and delivered lectures at many international conferences and workshops. He is a co-author for three textbooks, co-editor of three proceedings, and author of many research articles both in peer-reviewed journals and in proceedings.

Nilanjan Dey is an assistant professor in Department of Information Technology at Techno India College of Technology, Kolkata, India. He is a visiting fellow of the University of Reading, UK. He was an honorary Visiting Scientist at Global Biomedical Technologies Inc., CA (2012–2015). He was awarded his PhD from Jadavpur University in 2015. He is the editor-in-chief of *International Journal of Ambient Computing and Intelligence* and associate editor of *IEEE Access and International Journal of Information Technology*. He is the series co-editor of *Springer Tracts in Nature-Inspired Computing*, series co-editor of *Advances in Ubiquitous Sensing Applications for Healthcare*, series editor of *Computational Intelligence in Engineering Problem Solving and Intelligent Signal Processing and Data Analysis*. His main research interests include medical imaging, machine learning, computer-aided diagnosis, data mining, etc. He is the Indian Ambassador for the International Federation for Information Processing (IFIP)—Young ICT Group. He has been awarded as one among the top ten most-published academics in the field of computer science in India (2015–2017).

1

A Novel Stacked Model Ensemble for Improved TB Detection in Chest Radiographs

Sivaramakrishnan Rajaraman, Sema Candemir, Zhiyun Xue, Philip Alderson, George Thoma, and Sameer Antani

CONTENTS

1.1 Introduction

Tuberculosis (TB) is an infectious disease caused by a rod-shaped bacterium called *Mycobacterium tuberculosis*. According to the 2018 World Health Organization (WHO) report, there were an estimated 10 million new TB cases, but only 6.4 million (64%) were reported for treatment [1]. Countries including India, China, Pakistan, South Africa, and Nigeria accounted for more than 60% of the people suffering from the infection. A chest X-ray (CXR), also called chest film or chest radiograph, is the most common imaging modality used to diagnose conditions affecting the chest and its contents [2, 3]. CXR diagnosis has revolutionized the field of TB diagnostics and is extremely useful in establishing a plausible diagnosis of the infection.

Clinicians initiate treatment for the infection based on their judgment of these radiology reports. Posterior-anterior (PA) and lateral CXR projections are routinely examined to diagnose the conditions and provide diagnostic evidence [4]. Figure 1.1 (a)–(e) shows some instances of abnormal and normal CXRs.

With significant advancements in digital imaging technology there is an increase in the use of CXRs for TB screening. However, there is a lack of expertise in interpreting radiology images, especially in TB endemic regions, which adversely impacts screening efficacy [5], an ever-growing backlog and increased opportunity for disease spread. Also, studies show that there is a high degree of variability in the intra-reader and inter-reader agreement during the process of scoring CXRs [6]. Thus, current research is focused on developing cost-effective, computer-aided diagnosis (CADx) systems that can assist radiologists in interpreting CXRs and improve the quality of diagnostic imaging [7]. These systems are highly competent in reducing intra-reader/inter-reader variability and detection errors [8–11]. There are several prior approaches using traditional image analysis and machine learning (e.g. support vector machine [SVM]) that are valuable for providing background on CADx tools for CXR analysis [12–15]. The reader is referred to these as background. They are promoted as a convenient tool to be used in systematic screening and triaging algorithms due to the increased availability of digital radiography, which presents numerous benefits over conventional radiography, including enhanced image quality, safety, and reduced operating expenses [16]. CADx tools have gained immense significance; the appropriate use and advancing of these systems could improve detection accuracy and alleviate the human burden in screening. Earlier CADx studies were based on image segmentation and textural feature extraction with grey-level co-occurrence matrix [17]. A CADx system for TB detection was proposed by Van Ginneken et al. [18], who used multi-scale feature banks for feature extraction and a weighted nearest-neighbor classifier for classification of TB-positive and normal cases. The study demonstrated area under a curve (AUC) values of 0.986 and 0.82 on two private CXR datasets. A technique based on pixel-level textural abnormality detection was proposed by Hogeweg et al. [19] to obtain AUC values between 0.67 and 0.86. However, a

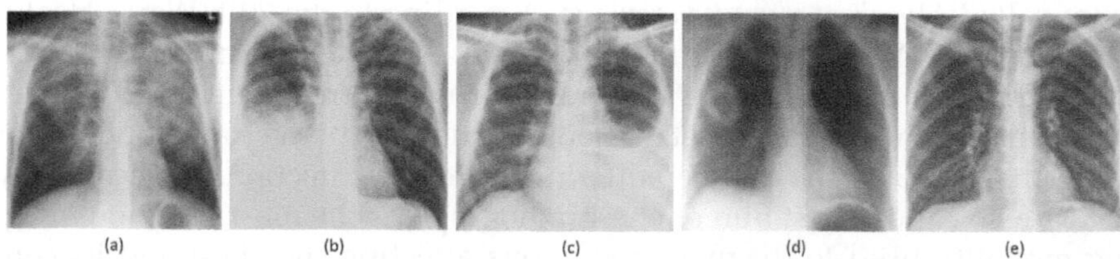

FIGURE 1.1
CXRs: (a) hyper-lucent cystic lesions in the upper lobes, (b) right pleural effusion, (c) left pleural effusion, (d) cavitary lung lesion in the right lung, and (e) normal lung.

comparative study of the proposed methods was hampered due to unavailability of public CXR datasets. Jaeger et al. [20] made available the public CXR datasets for TB detection, followed by Chauhan et al. [21], who helped to evaluate the proposed techniques on public datasets. Melendez et al. proposed the multiple instance learning methods for TB detection, which used moments of pixel intensities as features to be classified by an SVM classifier [22]. The authors obtained AUC between 0.86 and 0.91 by evaluating on three private CXR datasets. Jaeger et al. [5] proposed a combination of standard computer vision algorithms for extracting features from chest radiographs. The study segmented the region of interest (ROI) constituting the lungs, and extracted the features using a combination of algorithms that included a histogram of oriented gradients (HOG), local binary patterns (LBP), Tamura feature descriptors, and other algorithms. A binary classifier was trained on these extracted features to classify normal and TB-positive cases. CADx software based on machine learning (ML) approaches using a combination of textural and morphological features is also commercially available. This includes CAD4TB, a CADx software from the Image Analysis Group, Nijmegen, Netherlands that reported AUC ranging from 0.71 to 0.84 in a sequence of studies performed in detecting pulmonary abnormalities [23]. Another study achieved AUC of 0.87 to 0.90 by using an SVM classifier to classify pulmonary TB from the normal instances using texture and shape features [24]. However, the performance of textural features was found to be inconsistent across the imaging modalities. These features performed well as long as they were able to correlate with the disease, but delivered sub-optimal performance in instances when there was an overlapping of anatomical sites and images having complex appearances [25]. Feature descriptors such as bag-of-words (BOW) were also used in discriminating normal from pathological chest radiographs [26]. The method involves representing an image using a bag of visual words, constructed from a vocabulary of features extracted by local/global feature descriptors. A majority of CADx studies used handcrafted features that demand expertise in analyzing the images and account for variability in the morphology and texture of the ROI. On the other hand, deep learning (DL) models learn hierarchical layer-wise representation to model data at more and more abstract representations. These models are also known as hierarchical ML models that use a cascade of layers of non-linear processing units for end-to-end feature extraction and classification [27]. Convolutional neural networks (CNN), a class of DL models, have gained immense research prominence in tasks related to image classification, detection, and localization, as they deliver promising results without the need for manual feature selection [28]. Unlike kernel-based algorithms such as SVMs, DL models exhibit improved performance with an increasing number of training samples and computational resources [29].

Medical images contain visual representations of the interior of the body that aid in clinical analysis and medical intervention [30]. These images are specific to the internal structures of the body and have less in common with natural images. Under these circumstances, a customized CNN, specifically trained on the underlying biomedical imagery, could learn "task-specific" features to aid in improved accuracy. The parameters of a custom model could be optimized for improvement in performance. The learned features and salient network activations could be visualized to understand the strategy the model adapts to learn these task-specific features [31]. However, the performance improvement of customized CNNs comes at the cost of huge amounts of labeled data, which are difficult to obtain, particularly in biomedical applications. Transfer Learning (TL) methods are commonly used to relieve issues with data inadequacy where DL models are pre-trained on large-scale datasets [32]. These pre-trained models could be used either as an initialization for visual recognition tasks or as feature extractors from the underlying data [33]. There are several pre-trained CNNs available, including AlexNet [34], VGGNet [35], GoogLeNet [36], ResNet [37], etc., which transfer knowledge gained from learning a comprehensive feature set from the large-scale datasets to the underlying task and serve as feature extractors in an extensive range of visual recognition applications, outperforming the handcrafted features [38]. Study of the literature reveals the use of pre-trained CNNs in detecting pleural effusion and cardiomegaly in chest radiographs [39]. The performance of the pre-trained models was compared to that of the classifiers operated on handcrafted features, including LBP [40], GIST [41], and PiCo descriptors [42]. It was observed that the combination of pre-trained CNN and PiCo features gave the best performance with an AUC of 0.89 and 0.93 for cardiomegaly and right pleural effusion respectively. The first application of CNNs to TB detection was proposed by Hwang et al. [43], who customized the architecture of AlexNet and trained on a private CXR dataset. The results obtained with random weight initializations were not promising; however, the model performed better, with an accuracy of 0.77 and an AUC of 0.82, when trained with pre-trained weights. The authors also evaluated the performance with publicly available Montgomery and Shenzhen datasets [20] to obtain an accuracy of 0.674 and 0.837, respectively. The application of CNN toward TB detection was demonstrated in another study that used a custom CNN model, a variant of the AlexNet framework, trained on a private CXR dataset of approximately 10,000 images [43]. The study achieved unsatisfactory results with the custom model with random weight initializations. However, with pre-trained CNNs, better results were obtained on the Montgomery and Shenzhen CXR datasets, achieving AUC of 0.884 and 0.926 respectively. In another study, the authors assessed the accuracy and stability in findings of DL models toward detecting abnormalities on frontal CXRs. The de-identified radiographs were processed with the Qure AI tool [44]. The scores were generated, recorded, and compared with the standard of reference (SOR) established with expert radiologists' assessment toward detecting hilar prominence, pulmonary opacity, pleural effusion, and enlarged cardiac silhouette. It was observed that the value of

AUC ranged between 0.837–0.929 and 0.693–0.923 for DL and test radiologists, respectively. DL models had the lowest AUC of 0.758 toward assessing changes in pulmonary opacities. No statistical difference was observed between DL and SOR for these abnormalities. In a recent study, pre-trained CNNs were used as feature extractors for TB detection [45]. The study presents three different proposals for applying the pre-trained CNNs to extract features from the CXRs toward improving the accuracy of TB detection. The study used publicly available CXR datasets [20] and demonstrated superior performance to state-of-the-art methods.

The pioneering work on ensemble learning (EL) proved that multiple, diverse, and accurate base-learners could asymptotically build a strong-learner [46]. The generation of model ensembles could be categorized into homogeneous and heterogeneous methods. In the homogeneous method, base-learners use the same learning algorithm with different settings for training data and learning parameters. Some examples include the Bagging [47] and Boosting methods [48]. Heterogeneous methods generate base-learners with diverse learning algorithms. Different fusing strategies are used to combine the decision made by the base-learners. Majority voting [49] is a commonly used fusing method in which base-learners vote for a specific class and the predicted class collects the majority of votes. Simple average and weighted average methods [50] are also used. Stacking, otherwise called stacked generalization, is an optimal fusing technique that highlights each base-learner when it performs best and discredits it when it delivers sub-optimal performances [51]. Figure 1.2 gives a pictorial representation of this concept. This method introduces the concept of a second-level meta-learner that optimizes the combination of individual base-learners.

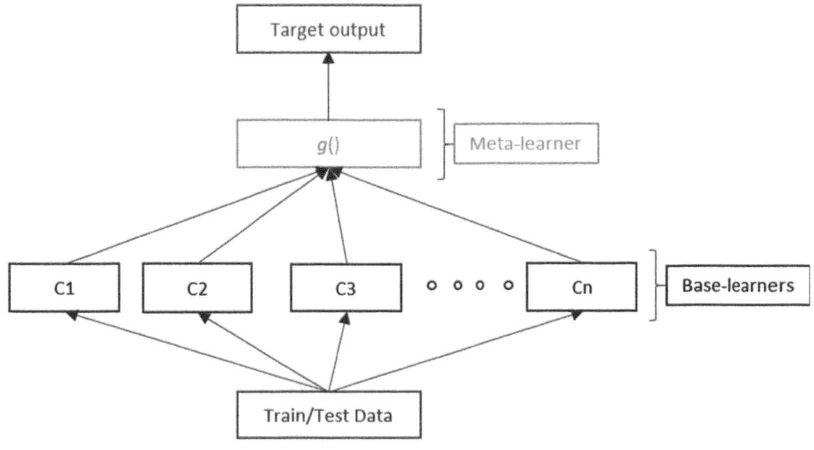

FIGURE 1.2
The concept of stacking.

Base-learners are diverse. They make different errors in the data and are accurate in different regions of the feature space. Individual base-learners are represented by C={C1, C2, ..., Cn} where "n" is the number of base-learners. The meta-learner g () learns how the individual base-learners make errors, and estimates and corrects their biases. Since both DL and EL have their inherent advantages in constructing non-linear decision-making functions, the combination of the two could efficiently handle tasks related to analyzing and understanding the underlying data. However, the literature is sparse in the use of ensemble models for TB detection. In the only study performed by Lakhani and Sundaram [52], the authors evaluated the efficacy of an ensemble of deep CNNs toward TB detection in chest radiographs. The study used publicly available CXR datasets, pre-trained, and untrained CNNs and performed ensembles on the best-performing models It revealed that the ensemble of pre-trained CNNs along with data augmentation performed better with a sensitivity of 97.3% in comparison to other models.

This study aims to evaluate the performance of a stacked model ensemble that optimally combines the predictions using handcrafted feature descriptors and customized and pre-trained CNNs, and presents four different proposals toward improving the accuracy of TB detection from PA CXR images. In the first proposal, we use local and global feature descriptors including GIST, HOG, and SURF to extract the features from chest radiographs. The extracted features are used to train an SVM classifier to classify the normal and abnormal CXRs. In the second proposal, we evaluate the performance of a customized CNN-based DL model that learns task-specific features toward classifying TB-positive and healthy controls. The proposed model is optimized for its architecture and hyper-parameters by performing Bayesian optimization in the process of minimizing the classification error. In the third proposal, we use four different pre-trained CNN models to extract features from the chest radiographs, and an SVM classifier is trained on these features to detect TB manifestations. In the fourth and final proposal, we perform a stacked ensemble of models from different proposals to evaluate their performance on disease detection. The contributions of this work are as follows: (a) comparative analysis of the performance of local/global feature descriptors including GIST, HOG, and SURF toward classifying TB-positive and healthy chest radiographs, (b) proposing customized CNN-based DL models, optimized for their architecture and hyper-parameters toward learning task-specific features, (c) visualizing the learned features and salient network activations in the customized model to understand the learning dynamics, (d) comparing the performance of pre-trained DL models as feature extractors for the underlying task, and (e) evaluating the performance of stacked model ensembles toward the task of improving the accuracy of TB detection. This chapter is organized as follows: Section 1.2 elaborates on the materials and methods, Section 1.3 discusses the results, and Section 1.4 concludes the chapter.

1.2 Materials and Methods

1.2.1 Data Collection and Preprocessing

This study is evaluated on four CXR datasets that include the two publicly available datasets from Montgomery County, Maryland, and Shenzhen, China, maintained by the National Library of Medicine (NLM) [20]. The CXRs in the Montgomery collection have pixel resolutions of either 4892×4020 or 4020×4892. The CXRs in the Shenzhen collection have resolutions of approximately 3000×3000 pixels. The Montgomery collection has 58 TB-positive cases and 80 healthy controls. The Shenzhen dataset has a total of 662 CXRs, which include 336 TB-positive CXRs and 326 healthy controls. Ground truth information for these datasets is available in the form of clinical findings, roughly annotating the abnormal locations in the CXR images. The acquisition and sharing of these datasets are exempted from National Institutes of Health (NIH) IRB review (#5357). The third dataset is from India, acquired by the National Institute of Tuberculosis and Respiratory Diseases, New Delhi, and made available by the authors [21]. This dataset contains two subsets of CXR collections from different X-ray machines and a balanced distribution of CXR images for TB-positive and normal cases. The CXRs in the India collection have resolutions ranging from 1024×1024 to 2480×2480 pixels. GT labels are available as global annotations for the normal and abnormal classes. TB manifestations in this dataset are obvious and distributed throughout the lungs. The fourth dataset is a private collection of CXRs obtained from Kenya under a retrospective study agreement with Moi Teaching and Referral Hospital, Eldoret, Kenya and with the assistance of Indiana University School of Medicine and Academic Model Providing Access to Healthcare (AMPATH), a Kenyan NGO. This dataset contains 238 abnormal CXRs and 729 healthy controls. Disease labels are made available as lung zone-based clinical findings from expert radiologists. The CXRs in the Kenya collection have resolutions of either 2004×2432 or 1932×2348 pixels.

The datasets include PA CXRs that contain regions other than the lungs which are irrelevant for lung TB detection. To alleviate issues due to models learning features that are irrelevant to detecting lung TB and demonstrate sub-optimal performance, the lung region constituting the ROI is segmented by a method that uses anatomical atlases with non-rigid registration [24]. This segmentation method follows a content-based image retrieval approach to identify the training examples that bear resemblance to the patient CXR by using Bhattacharyya similarity measure and partial Radon transform. The patient-specific anatomical lung shape model is created using SIFT-flow [53] for registering the training masks for the patient CXRs. The refined lung boundaries are extracted using graph-cut optimization and customized energy function [54]. An instance of a CXR with the detected lung region and cropped lung area using the proposed method is shown in Figure 1.3.

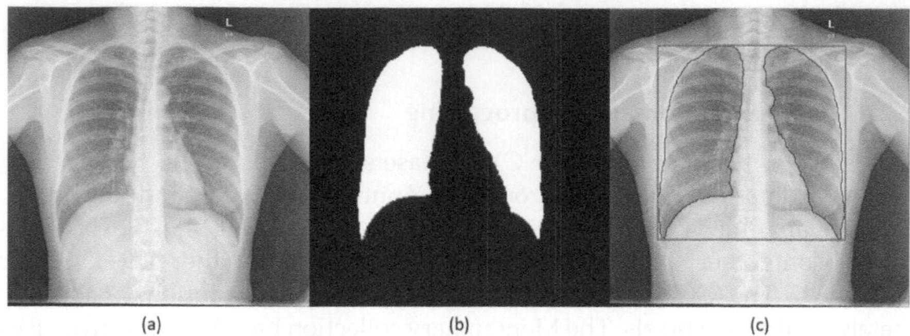

FIGURE 1.3
Lung ROI segmentation: (a) CXR, (b) computed lung mask, (c) Segmented ROI.

After lung segmentation, the resulting image is cropped to the size of a bounding box that contains all the lung pixels. The resultant images are enhanced for contrast by applying contrast limited adaptive histogram equalization (CLAHE).

1.2.2 Proposal 1—Feature Extraction Using Local/Global Feature Descriptors and Classification Using SVM

In the first proposal (P1), we evaluated the performance of global descriptors including GIST and HOG and local descriptors including SURF toward identifying TB manifestations. Since pre-trained CNNs demand down-sampling of the underlying data to fit the specific requirements of the input layer, a lot of potentially viable information pertaining to the signs of TB infection may be lost. The best way to overcome this issue is to use the local/global feature descriptors to extract discriminative information from the entire CXR image without the need for rigorous down-sampling. The GIST feature descriptor summarizes the information pertaining to the gradients, orientations, and scales for different regions of a given image to provide a robust description of the image. The process results in image filtering into low-level features including intensity, color, motion, and orientation at multiple scales in space. GIST captures these features toward identifying the salient image locations that significantly differ from those of the neighbors. Given an input image, the GIST descriptor convolves the image with 32 Gabor filters [55] at four different scales and eight different orientations, producing a total of 32 feature maps with the same size as that of the input image. The process results in the computation of the direction of low- and high-frequency repetitive gradients for a given image. Each of these feature maps is then divided into 16 regions with a 4×4 square grid and the feature values are averaged within each sub-region. The averaged values from the 16 sub-regions are concatenated for the 32 different feature maps, resulting in a total of 512 GIST descriptors for a given image. Across the datasets, eight orientations per scale and four blocks are used in this study.

HOG feature descriptors were introduced by Dalal and Triggs [56], and are used in computer vision applications for object-detection tasks for the purpose of counting the gradient orientation occurrences in localized image regions. HOG measures the first-order image gradient pooled in overlapping orientation bins, and gives a compressed and encoded version of an image. It counts the occurrences of different gradient orientations and maintains geometric invariance and photometric transformations. The process involves computing the gradients, creating cell histograms, and generating and normalizing the descriptor blocks. Given an image, HOG computes the gradient orientations and plots the histogram of these orientations, giving the probability of the existence of a given gradient with a specific orientation in a given path. The features are extracted over small blocks in a repetitive fashion to preserve information pertaining to the local structures and the block-wise features are finally concatenated into a feature vector. HOG descriptors are computed on a dense grid of uniformly spaced cells and use overlapping local contrast normalization to aid in improved accuracy. An increase in the cell size helps to capture spatial information on a large scale. A block comprises a number of cells, and a reduced block size helps to capture the significance of local pixels and suppress changes in illumination. In this proposal, the number of cells overlapping between adjacent blocks is chosen to be half the block size to ensure adequate normalization of contrast. The cell size is varied and the results are visualized across the datasets to observe the degree of variation in the amount of shape information encoded in the feature vector. We also visualized the effect of a reduced block size in the process of capturing the significance of local pixels and suppressing changes due to illumination variations. We empirically evaluated the values for the cell size parameter, number of bins, and block size that gave the best accuracy and set them to [32 32], 9, and [2 2] respectively.

BOW is a technique adapted from the world of information retrieval to computer vision applications. Contrary to text, images do not contain words. So, this method creates a bag of features extracted from the images across the classes, using a custom feature descriptor, and constructs a visual vocabulary. In this study, speeded-up robust features (SURF) are used as feature descriptors that detect interesting key points in a given image by using an integer approximation of the determinant of a blob detector based on the Hessian matrix [57]. The feature point locations across the CXR images are selected through a grid method and SURF are extracted from the selected locations. A grid step is chosen and the features are extracted from the CXR images across the normal and TB-positive categories. The number of features across the image categories is balanced to improve clustering. A visual vocabulary is created by reducing the dimension of the features through feature space quantization, using K-means clustering. Images are encoded into feature vectors and the encoded training samples across the image categories are fed into the SVM classifier to be classified into TB-positive and normal categories. An [8 8] grid step and 500 clusters are used in this study

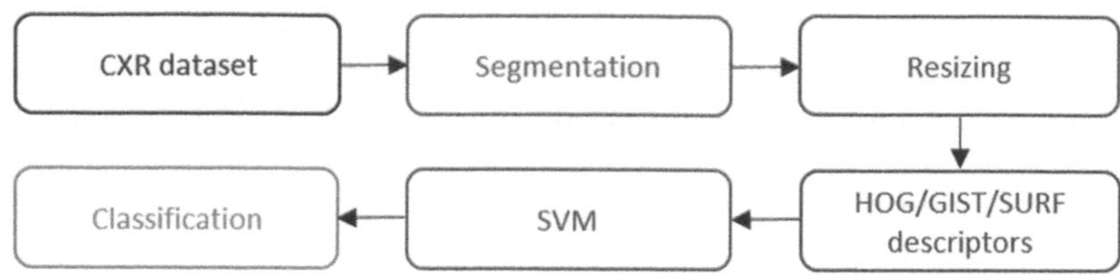

FIGURE 1.4
Steps involved in Proposal 1.

across the datasets. For this proposal, we performed a nested cross-validation. In the outer loop, we performed five-fold cross-validation for all the datasets. In the inner loop, we performed Bayesian optimization [58] to minimize classification error by varying the parameters for the SVM classifier which includes the box constraint, kernel scale, kernel function, and order of the polynomial. The chosen ranges include [1e-3 1e3], [1e-3 1e3], and [2 4] for box constraint, kernel scale, and order of the polynomial respectively. For the kernel function, the optimization process searched among linear, Gaussian, RBF, and polynomial kernels. Figure 1.4 shows the steps involved in this proposal.

1.2.3 Proposal 2—Feature Extraction and Classification Using a Customized CNN

In the second proposal (P2), we evaluated customized CNN models toward the task of TB detection. As stated earlier, we were interested in optimizing a customized model to learn task-specific features. Train/validation splits are randomized (70/30). Images are down-sampled to 224 × 224 pixel resolutions and training samples are augmented with horizontal and vertical translations in the range of [–5 5] pixels and rotations in the range of degrees [–10 10], toward preventing model overfitting. We made sure to augment only the training data to suit the deployment scenario where abrupt mirroring, flipping, and a huge degree of rotations are not viable. Figure 1.5 shows the steps involved in this proposal. We applied Bayesian optimization [58] to find the optimal network parameters and training options for the custom CNNs trained on different datasets. Bayesian optimization is applied to optimize

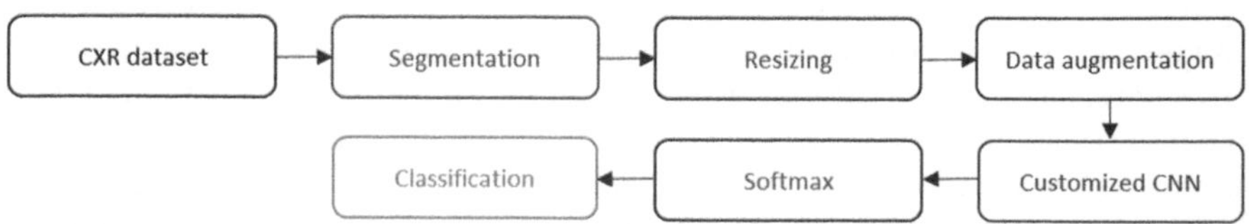

FIGURE 1.5
Steps involved in Proposal 2.

non-differentiable, discontinuous functions by maintaining a Gaussian process model of the objective function to be minimized and to perform objective function evaluations for training to find the optimal model parameters for the underlying data.

The framework for the customized CNN is specified. Each CNN block has a convolutional layer, followed by batch normalization [59], and Rectified Linear Units (ReLU) layer [34]. Padding is added to the convolutional layers to ensure that the spatial output dimensions match the original input. The number of filters is increased by a factor of two, and every time a max-pooling layer is used to ensure the amount of computation roughly remains the same across the convolutional layers. A filter size of 3×3 is used uniformly across the layers. The number of filters in a given layer is chosen to be $1/\sqrt{\text{network depth}}$ so that the customized CNNs with different network depths have roughly the same number of parameters and demand the same computational cost per iteration. The initial number of filters in each convolutional layer is chosen to be $\left(\text{round}\left(\text{image size}/\sqrt{\text{network depth}}\right)\right)$. The variables to be optimized are chosen and search ranges are specified. These ranges include [1 3], [1e–3 5e–2], [0.8 0.99], and [1e–10 1e–2] for the network depth, learning rate, stochastic gradient descent (SGD) momentum, and L2-regularization parameters respectively. An objective function for the Bayesian optimization process that takes, as its inputs, the values of the optimization variables, is used for training the customized CNN across different datasets, and a classification error is returned. Bayesian-optimized parameters for the least classification error are recorded.

1.2.4 Proposal 3—Feature Extraction Using Pre-Trained CNNs and Classification Using SVM

In the third proposal (P3), we evaluated the performance of pre-trained CNNs as feature extractors toward classifying TB-positive and healthy CXR images. Figure 1.6 shows the steps involved in this proposal. We evaluated state-of-the-art CNN models including AlexNet, VGG-16, GoogLeNet, and ResNet-50. The segmented ROI constituting the lungs are down-sampled to match the input dimensions of the pre-trained models. Each layer of the

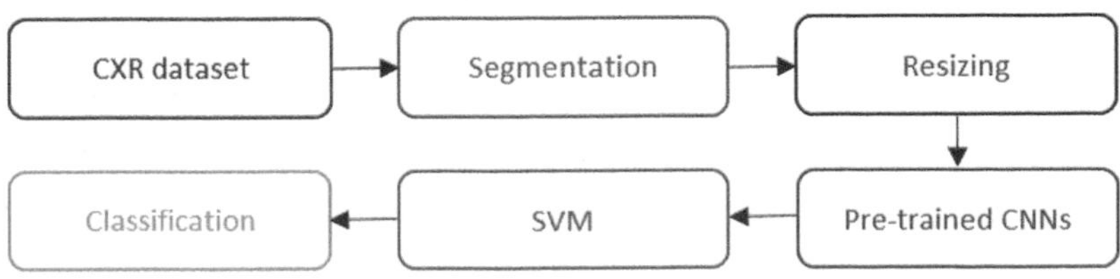

FIGURE 1.6
Steps involved in Proposal 3.

pre-trained CNNs produces an activation for the given image. Earlier layers capture primitive features that include blobs, edges, and colors which are abstracted by the deeper layers to form higher-level features to present a more affluent image representation. These features are extracted from the layer before the classification layer [33] and used to train an SVM classifier. Figure 1.6 shows the steps involved in this proposal. As in P1, we performed a nested cross-validation. In the inner loop, we performed Bayesian optimization to minimize the cross-validation error by varying the SVM parameters. The chosen ranges include [1e v–3 1e3], [1e–3 1e3], and [2 4] for box constraint, kernel scale, and order of the polynomial respectively. For the kernel function, the optimization process searched among linear, Gaussian, RBF, and polynomial kernels.

1.2.5 Proposal 4—Constructing Stacked Model Ensembles

Literature reveals the usage of local/global feature descriptors, and customized and pre-trained CNNs in classifying medical images. However, there are no state-of-the-art methods available that evaluate the performance of a stacked generalization of these approaches toward TB detection. It has been shown that, in general, a classifier's performance could be improved using a stacked model ensemble that combines multiple, diverse base-learners which make independent errors via a meta-learner [51]. These base-learners use different learning algorithms to overfit different regions in the feature space; thus, the stacked ensemble is often heterogeneous. In the fourth and final proposal (P4), we created a stacked generalization of models from different proposals to find an optimal stacked ensemble model that improved the accuracy of TB detection. Stacked ensembles are created for the models in P1 (E [P1]), P1 and P2 (E [P1, P2]), P1 and P3 (E [P1, P3]), P2 and P3 (E [P2, P3]), and P1, P2, and P3 (E [P1, P2, P3]). Figure 1.7 shows the steps involved in this proposal.

Figure 1.8 shows the schematic of the stacked ensemble of models from different proposals. Stacked model ensembles are implemented with the best parameter values for the models from different proposals and evaluated through five-fold cross-validation. The model ensembles consist of two levels. At the first level (Level–0), models from different proposals

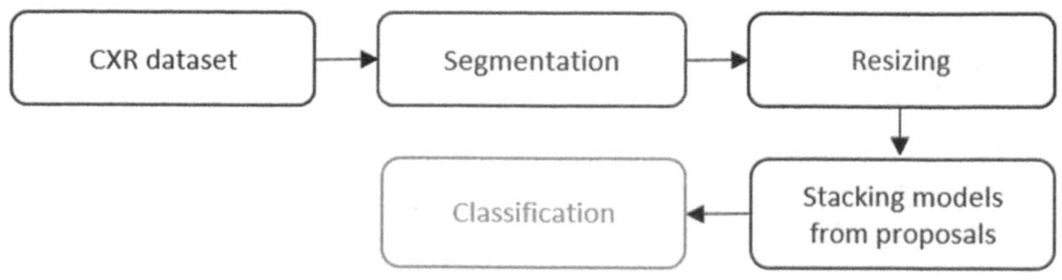

FIGURE 1.7
Steps involved in Proposal 4.

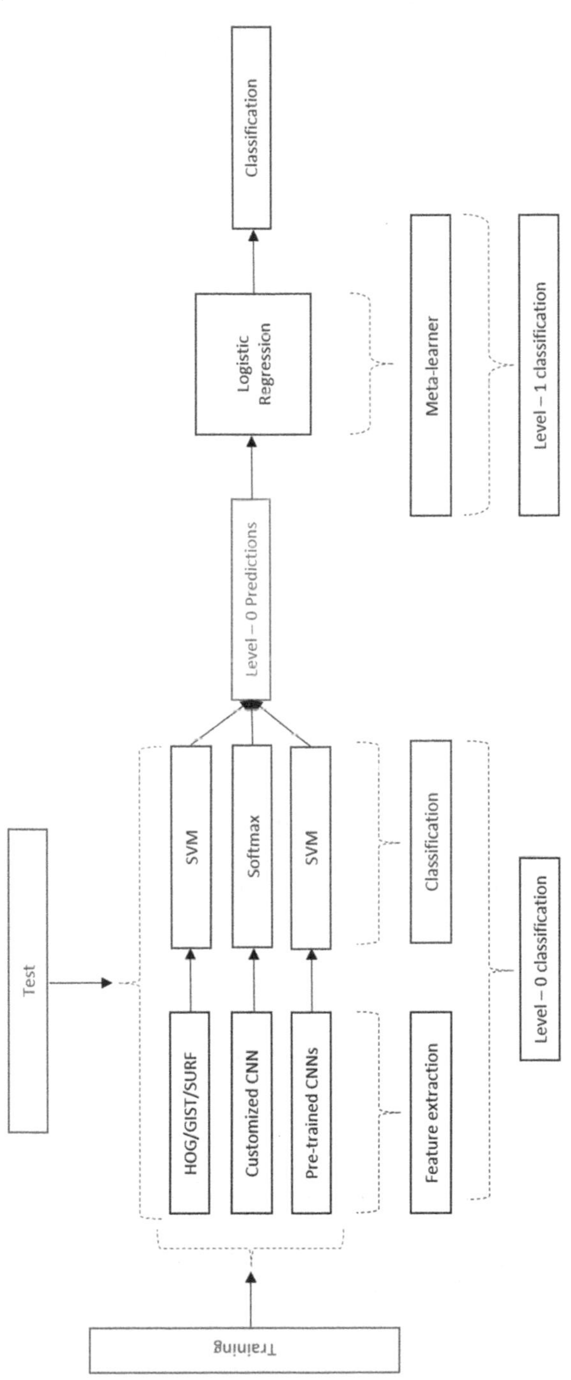

FIGURE 1.8
Stacked generalization of models from different proposals.

constitute diverse, individual base-learners. Let M denote the number of rows in the training data and N denote the number of base-learners. An M×N matrix, along with the original responses for the training samples, constitutes Level–0 predictions. The meta-learner is trained on the predictions of the base-leaners. The meta-learner is a logistic regression classifier that estimates the probability for a binary response. The stacked ensemble of base-learners and meta-learner is used to predict on the test data. In this study, we used a Windows® system with Intel® Xeon® CPU E5-2640v3 2.60-GHz processor, 1 TB of Hard Disk space, 16 GB RAM, a CUDA-enabled Nvidia® GTX 1080 Ti 11GB graphical processing unit (GPU), Matlab® R2017b, Weka® 3 ML software, and CUDA 8.0/cuDNN 5.1 dependencies for GPU acceleration.

1.3 Results and Discussion

In P1, we used global descriptors including GIST and HOG, and local descriptors including SURF, toward identifying TB manifestations. The images are down-sampled to 3072×3072, 4096×4096, 2048×2048, and 1024×1024 pixel resolutions for Shenzhen, Montgomery, Kenya, and India collections respectively. Table 1.1 shows the results obtained with different feature descriptors in terms of accuracy and AUC. For the Shenzhen dataset, the best results are obtained with the GIST features and SVM/RBF, with an accuracy of 0.845 and AUC of 0.921. For the Montgomery dataset, the BOW model using SURF and SVM/RBF demonstrated superior performance with an accuracy of 0.775 and AUC of 0.845. For the Kenya dataset, HOG features and SVM/Gaussian showed better performance in terms of accuracy, but the GIST features and RBF kernel-based SVM classifier gave the best AUC of 0.748. For the India dataset, GIST features and SVM/RBF demonstrated superior performance with an accuracy of 0.882 and AUC of 0.961. We could observe that no feature descriptor performed equally well on the underlying data. The chest radiographic images across the

TABLE 1.1

P1— GIST, HOG, and SURF-Based Feature Extraction and SVM-Based Classification

	Accuracy			AUC		
Datasets	**HOG**	**GIST**	**SURF**	**HOG**	**GIST**	**SURF**
Shenzhen	0.841	**0.845**	0.816	0.917	**0.921**	0.890
Montgomery	0.708	0.750	**0.775**	0.772	0.817	**0.845**
Kenya	**0.683**	0.667	0.672	0.741	**0.748**	0.747
India	0.880	**0.882**	0.864	0.947	**0.961**	0.938

datasets are collected with different machinery and at different pixel resolutions. The local/global feature descriptors are rule-based feature extraction mechanisms that are built and optimized to improve performance on individual datasets. For this reason, they don't perform equally well across the datasets. It can be noted here that the results obtained with the India dataset are superior to those obtained with the other datasets. A similar pattern is observed in the result tables for different proposals. A noteworthy factor with the India dataset is that though the dataset is limited, TB manifestations are obvious and are distributed throughout the lungs, which gives the feature descriptors the opportunity to capture highly discriminative features across normal and abnormal categories. The lowest performance is observed with the Kenya dataset, the principal reason being that the dataset has a highly imbalanced distribution of instances across the classes, with 238 abnormal CXRs in comparison to 729 healthy controls. The patients were all HIV+ with a low immune response. For this reason, the expression of the disease, even in severe cases, is significantly weaker than with ordinary TB. Also, the CXRs are obtained as a result of mobile truck-based screening, the images are cassette-based, and hence the image resolution is not commendable even after CLAHE enhancement, which further impaired the performance of feature extraction and classification. With the Montgomery dataset, performance limitation may be attributed to the limited size of the dataset and also to the degree of imbalance across the classes, where 40% of the samples are TB-positive as compared to 60% of the healthy controls.

In P2, we evaluated a customized CNN model for each dataset, optimized to learn task-specific features toward the task of TB detection. The images are down-sampled to 224×224 pixel resolution. The architecture of the optimized customized CNN models is shown in Figure 1.9. We applied Bayesian optimization to find the optimal network parameters and training options for the customized models. Table 1.2 shows the optimal values for the parameters learned by the customized CNN for the different datasets. Table 1.3 presents the performance measures of the customized CNNs across the datasets. The performance of the customized CNN, with respect to the India dataset, is superior to that with the other datasets, obtaining an accuracy of 0.860 and AUC of 0.937. The custom model showed the least performance with the Kenya dataset for the reasons already discussed. Also, the customized CNN is not able to completely reap the benefits of task-specific learning and classification due to data scarcity. The training samples are augmented only to resolve overfitting concerns, but this did not improve the validation accuracy.

We visualized the task-specific features and salient network activations in the optimized CNN model towards understanding the learning dynamics. For instance, we took the optimized CNN model for the India dataset and visualized the features in the second, fourth, sixth, and ninth convolutional layers, as shown in Figure 1.10 (a)–(g). The convolutional layers outputs

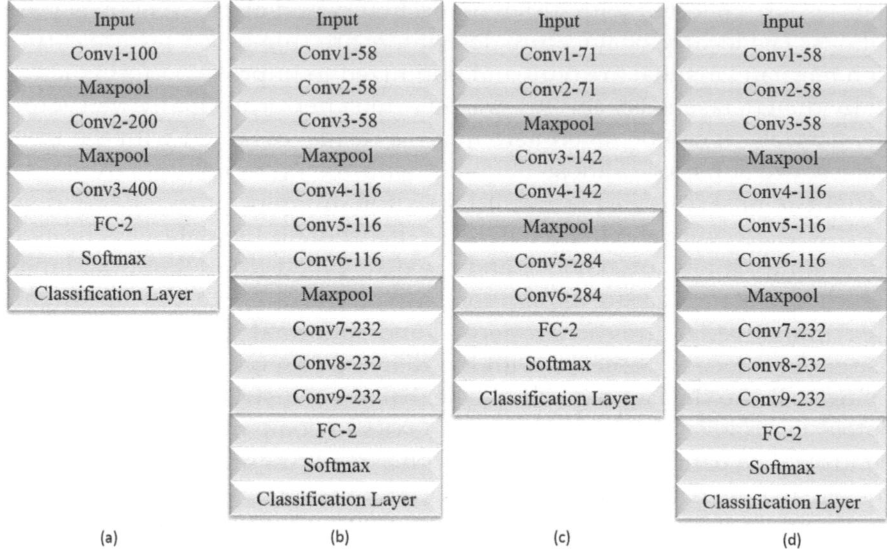

FIGURE 1.9
Optimized CNNs for different datasets: (a) Shenzhen, (b) Montgomery, (c) Kenya, (d) India.

TABLE 1.2
Optimal Parameters for the Customized CNN Model

	Parameters				Performance	
Datasets	**Momentum**	**Learning Rate**	**L2-Decay**	**Network Depth**	**Accuracy**	**AUC**
Shenzhen	0.841	3.615e-04	1.608e-10	3	0.820	0.894
Montgomery	0.841	9.353e-04	1.448e-08	9	0.750	0.817
Kenya	0.845	3.949e-04	2.058e-09	6	0.698	0.761
India	0.942	1.994e-04	3.600e-03	9	**0.860**	**0.937**

TABLE 1.3
P2— Customized DL Model-Based Feature Extraction and Classification

Datasets	Accuracy	AUC
Shenzhen	0.820	0.894
Montgomery	0.750	0.817
Kenya	0.698	0.761
India	0.860	0.937

multiple channels, each corresponding to a filter applied to the input layer. The fully connected layer abstracts the features from the earlier layers and outputs channels corresponding to the image categories. The second convolutional layer appears to learn mostly colors and edges, indicating that the channels

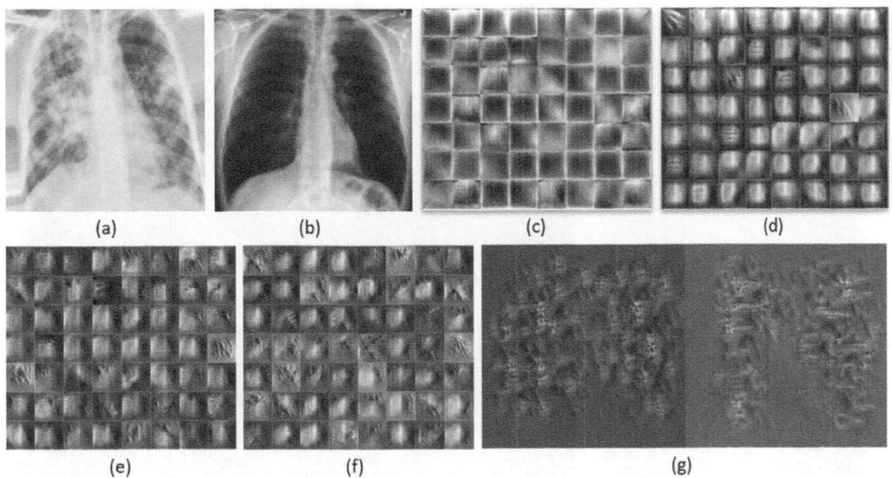

FIGURE 1.10
Visualizing features. From left to right: (a) abnormal CXR, (b) normal CXR, (c) second convolution layer, (d) fourth convolution layer, (e) sixth convolution layer, (f) ninth convolution layer, (g) fully connected layer.

are color filters and edge detectors. As we progressed to the fourth convolutional layer, we observed that the customized CNN began to learn the edges and orientations. As we progressed to the sixth and ninth convolutional layers, we visualized task-specific features leading to the formation of shapes by abstracting the primitive features from the earlier layers. The fully connected layer towards the end of the model loosely resembled the abnormal and normal classes respectively. The model understands a decomposition of its visual input space as a hierarchical-modular network of filters and a probabilistic mapping between combinations of filters and a set of labels. This is not analogous to "seeing," as programmed in the visual cortices of humans. The exact nature of the convolutional filters, the hierarchy, and the process through which they learn have very little in common with the human visual cortices, which are not convolutional and are structured into cortical columns whose purpose is yet to be fully explored. The visual perception of humans is sequential, active, and involves motor control; thus it is very different from that of the convolutional filters.

We also visualized the salient network activations to discover the learned features by comparing the areas of activation against the original image. An abnormal CXR is fed into the optimized CNN model and the activations of different layers are analyzed. The performance of the model is evaluated by investigating the activation of the channels on a set of input images. The resulting activations are compared with that of the original image, as shown in Figure 1.11 (a)–(d). CNN learns to detect complicated features in deeper convolutional layers that build up their features by combining features from the earlier layers. Thus, the channels of the deepest convolutional layer in the model are analyzed to

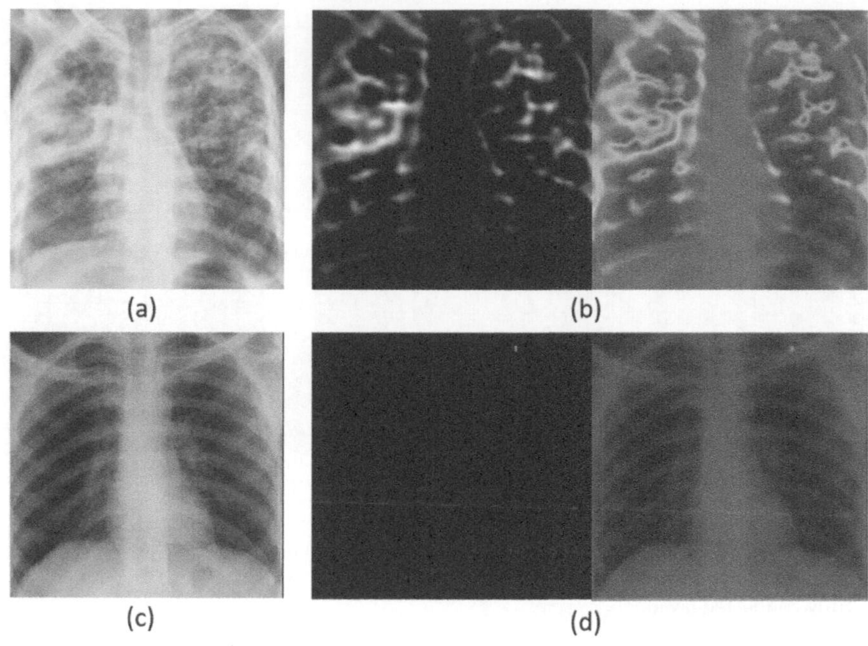

FIGURE 1.11
Visualizing the highest channel activations in the deepest convolutional layer: (a) CXR showing bilateral pulmonary TB, (b) activations/heat maps, (c) normal lung, (d) activations/heat maps.

observe the areas activating on the image and compared to the corresponding areas in the original image. The channel showing the highest activation for the abnormal locations in the input images is investigated. All activations are scaled to the range [0, 1]. When the activations are strongly positive, they are observed as white pixels, and when strongly negative, by black pixels. When the activations are not strong enough, they are observed as gray pixels. The position of a pixel in the channel activation corresponds to that in the original image. The channels show both positive and negative activations. However, only positive activations are investigated because of the ReLU non-linearity following the convolutional layers. The activations of the ReLU layer show the location of abnormalities. CXRs showing bilateral pulmonary TB and normal lung are input to the model, and the saliency maps and the grayscale image showing the network channel activations are extracted. A pseudo-color image is generated to get a clearer and more appealing representation from the perceptual aspect by using the "jet" colormap, so that the activations higher than a given threshold appear bright red, with discrete color transitions in between. The threshold is selected to match the range of activations and achieve the best visualization effect. The resulting heat maps are overlaid onto the original image, and the black pixels in the heat maps are made fully transparent. It is observed from the heat maps that the customized model precisely activated the location of abnormalities and showed no activation for the normal lung image. This implies that the customized model learns task-specific features and localizes the abnormalities precisely to help distinguish between normal and abnormal classes.

Table 1.4 presents the results of P3 using pre-trained CNNs for feature extraction and SVM/RBF-based classification. Images are down-sampled to 224×224 and 227×227 pixel resolutions to suit the input requirements for the different pre-trained CNNs, across all the datasets. For the Shenzhen dataset, AlexNet obtained the best accuracy of 0.859 and AUC of 0.924. The same pattern is observed across Montgomery, Kenya, and India datasets. For the Montgomery dataset, AlexNet obtained the best accuracy of 0.725 and AUC of 0.817. For the India dataset, AlexNet outperformed the other pre-trained CNNs with an accuracy of 0.872 and AUC of 0.950. Only for the Kenya dataset did we observe that the AUC of VGG-16 is slightly better than that of AlexNet; however, the accuracy of AlexNet was higher than that of the other pre-trained CNNs. It can be noted that the results obtained with the India dataset are superior to the results obtained with the other datasets for the reasons discussed earlier. Among the pre-trained CNNs evaluated in this study, AlexNet outperformed the other models across the datasets. The deeper layers of ResNet-50 and GoogLeNet are progressively more complex, specific to the ImageNet dataset, and not suitable for the underlying task of binary medical image classification. For large-scale datasets such as ImageNet, deeper networks outperform shallow counterparts for the reason that the data is diverse and the networks learn abstractions for a huge selection of classes. In our case, for the binary task of TB detection, the variability in data is several orders of magnitude smaller and deeper networks do not seem to be a fitting tool. Also, literature studies reveal that convolutional features from shallow networks lead to higher accuracy than do the features of the deeper networks. Shallow models such as AlexNet provided high accuracy in the detection task [60]. Also, the top layers of pre-trained CNNs such as GoogLeNet and ResNet-50 are probably too specialized, progressively more complex, and not the best candidate to re-use for the task of our interest. This explains the difference in performance in our case.

Table 1.5 shows the results of the final proposal using ensembles of models from different proposals. The results obtained are promising in comparison

TABLE 1.4

P3—Pre-Trained CNNs-Based Feature Extraction and SVM-Based Classification

	Accuracy				AUC			
Datasets	**AlexNet**	**VGG-16**	**GoogLeNet**	**ResNet-50**	**AlexNet**	**VGG-16**	**GoogLeNet**	**ResNet-50**
Shenzhen	**0.859**	0.829	0.768	0.819	**0.924**	0.901	0.870	0.893
Montgomery	**0.725**	0.717	0.678	0.676	**0.817**	0.757	0.648	0.616
Kenya	**0.693**	0.691	0.674	0.678	0.776	**0.777**	0.750	0.753
India	**0.872**	0.812	0.796	0.812	**0.950**	0.892	0.888	0.902

Note: Bold numerical values indicate the performance measures of the best-performing model/s.

to that obtained from the other proposals. The stacked ensemble of local/global features (E[P1]) had similar accuracy values of 0.875 and 0.960 across the selection of ensembles for the Montgomery and India datasets. However, the ensemble of all the proposals (E[P1P2P3]) had the highest AUC of 0.986 and 0.995 in comparison to other stacked ensembles for the Montgomery and India datasets. The results are superior to those obtained from the other individual proposals. The same pattern is observed across all the datasets. One of the most significant requirements for the creation of stacked ensembles is that the base-learners be accurate and make diverse errors; i.e. the errors must have a low correlation [61]. Since we have a collection of models from different proposals, we benefit from the fact that their outputs are diverse and accurate, with less correlation in their errors, thus enhancing the performance of the ensembles. This is exactly what happened, as observed from the results.

Tables 1.6 and 1.7 compare the results obtained across the ensembles of different proposals presented in this study and literature on TB detection. The stacked ensemble results are, in almost all the cases, superior to those from different proposals. For this reason, the stacked ensemble of models from all the proposals (E[P1P2P3]) outperformed the other ensembles under study. In terms of accuracy, as shown in Table 1.6, the stacked ensemble of models from all the proposals (E[P1P2P3]) outperformed the state-of-the-art. The proposed ensemble demonstrated the highest accuracy of 0.960 for India, followed by 0.959 for the Shenzhen, 0.875 for the Montgomery, and 0.810 for

TABLE 1.5

P4—Ensemble of Models from Different Proposals

	E[P1]		E[P1P2]		E[P1P3]		E[P2P3]		E[P1P2P3]	
Datasets	**Accuracy**	**AUC**	**Accuracy**	**AUC**	**Accuracy**	**AUC**	**Accuracy**	**AUC**	**Accuracy**	**AUC**
Shenzhen	0.934	0.955	0.944	0.980	0.934	0.991	0.944	0.78	**0.959**	**0.994**
Montgomery	**0.875**	0.875	**0.875**	0.927	**0.875**	0.962	0.708	0.927	**0.875**	**0.986**
Kenya	0.733	0.825	0.784	0.826	0.776	0.826	0.767	0.765	**0.810**	**0.829**
India	**0.960**	0.960	0.940	0.958	**0.960**	0.965	0.940	0.974	**0.960**	**0.995**

Note: Bold numerical values indicate the performance measures of the best-performing model/s.

TABLE 1.6

Comparison with the Literature—Accuracy

	Literature				Proposed Approaches				
Datasets	**[5]**	**[43]**	**[45]**	**[21]**	**E[P1]**	**E[P1P2]**	**E[P1P3]**	**E[P2P3]**	**E[P1P2P3]**
Shenzhen	0.840	0.837	0.847	–	0.934	0.944	0.934	0.944	**0.959**
Montgomery	0.783	0.674	0.826	–	**0.875**	**0.875**	**0.875**	0.708	**0.875**
Kenya	–	–	–	–	0.733	0.784	0.776	0.767	**0.810**
India	–	–	–	0.943	**0.960**	0.94	**0.960**	0.940	**0.960**

Note: Bold numerical values indicate the performance measures of the best-performing ensemble/s.

TABLE 1.7

Comparison with the Literature—AUC

	Literature				Proposed Approaches				
Datasets	**[5]**	**[43]**	**[45]**	**[21]**	**E[P1]**	**E[P1P2]**	**E[P1P3]**	**E[P2P3]**	**E[P1P2P3]**
Shenzhen	0.900	0.926	0.926	–	0.955	0.98	0.991	0.780	**0.994**
Montgomery	0.869	0.884	0.926	–	0.875	0.927	0.962	0.927	**0.986**
Kenya	–	–	–	–	0.825	0.826	0.826	0.765	**0.829**
India	–	–	–	0.960	0.960	0.958	0.965	0.974	**0.995**

Note: Bold numerical values indicate the performance measures of the best-performing ensemble/s.

the Kenya datasets. We could observe similar patterns with the AUC values as shown in Table 1.7, where (E[P1P2P3]) clearly demonstrated a high AUC value in comparison to the results discussed in the literature. The results for the India dataset are superior, with an AUC of 0.995, followed by 0.994 for the Shenzhen, 0.986 for the Montgomery, and 0.829 for the Kenya datasets. As observed from these result tables, (E [P1P2P3]) achieved superior results across all the datasets.

1.4 Conclusion and Future Work

We have discussed four different proposals for improving the performance of TB detection. In P1, we used local and global feature descriptors to extract discriminative features from the radiographic images. The extracted features are used to train an SVM classifier. In P2, we optimized the architecture and hyper-parameters of customized CNNs using Bayesian optimization toward learning task-specific features for the underlying data. The customized model is highly compact, and has fewer trainable parameters and less architectural flexibility. The learned features and salient network activations are visualized to understand the learning dynamics. In P3, we used four different pre-trained CNNs to extract features from the datasets, and trained an SVM classifier on the extracted features. Under circumstances when the data availability is sparse, it is not recommended to fine-tune the pre-trained CNNs due to overfitting concerns. Literature studies [30] reveal that pre-trained CNNs could be used as a promising feature-extraction tool, especially for biomedical imagery. In P4, we performed a stacked ensemble of different proposals to find the optimal ensemble model that improved the accuracy of TB detection. Model stacking optimizes the combination of several diverse and accurate base-learners and reduces generalization errors. From our current studies, we believe that the stacked ensemble of diverse and accurate models using local/global feature descriptors and customized and pre-trained CNN models could be a promising option for improving the detection accuracy, particularly when the data are sparse. An appealing use case is to apply this method in applications with sparse data, particularly in biomedical imagery where the usage of only CNNs would lead to overfitting. The proposed ensemble could serve as triage, to minimize patient loss and reduce delays in resource-constrained settings; it could also be adapted to improving the accuracy of screening for other health-related applications. With regard to advancements in TB detection, recent works [62] demonstrate that the future demands large-scale biomedical datasets. The performance of the stacked ensemble could be highly promising with such a large collection of data.

Acknowledgments

This work was supported by the Intramural Research Program of the Lister Hill National Center for Biomedical Communications (LHNCBC), the National Library of Medicine (NLM), and the U.S. National Institutes of Health (NIH).

Conflict of Interest

The authors have no conflict of interest to report.

References

1. "WHO Global Tuberculosis Report 2018," 2018, Available at: https://www.who.int/tb/publications/global_report/en/, Accessed on 05/15/2018..
2. F. A. Mettler, W. Huda, T. T. Yoshizumi, and M. Mahesh, "Effective doses in radiology and diagnostic nuclear medicine: a catalog," *Radiology*, vol. 248, no. 1, pp. 254–263, 2008.
3. S. Rajaraman, S. Candemir, I. Kim, G. Thoma, S. Antani, "Visualization and interpretation of convolutional neural network predictions in detecting pneumonia in pediatric chest radiographs," *Applied Sciences*, vol. 8, no. 10, p. 1715, 2018.
4. M. F. Iademarco, J. O'Grady, and K. Lönnroth, "Chest radiography for tuberculosis screening is back on the agenda," *International Journal of Tuberculosis and Lung Disease*, vol. 16, no. 11, pp. 1421–1422, 2012.

5. S. Jaeger, A. Karargyris, S. Candemir, L. Folio, J. Siegelman, F. Callaghan, Z. Xue, K. Palaniappan, R. K. Singh, S. Antani, G. Thoma, Y. X. Wang, P. Lu, and C. J. McDonald, "Automatic tuberculosis screening using chest radiographs," IEEE Transactions on Medical Imaging, vol. 33, no. 2, pp. 233–245, 2014.
6. Y. Balabanova, R. Coker, I. Fedorin, S. Zakharova, S. Plavinskij, N. Krukov, R. Atun, and F. Drobniewski, "Variability in interpretation of chest radiographs among Russian clinicians and implications for screening programmes: observational study," *BMJ*, vol. 331, no. 7513, pp. 379–382, 2005.
7. S. Jaeger, A. Karargyris, S. Candemir, J. Siegelman, L. Folio, S. Antani, and G. Thoma, "Automatic screening for tuberculosis in chest radiographs: a survey," *Quantitative Imaging in Medicine and Surgery*, vol. 3, no. 2, pp. 89–99, 2013.
8. P. Maduskar, M. Muyoyeta, H. Ayles, L. Hogeweg, L. Peters-Bax, and B. van Ginneken, "Detection of tuberculosis using digital chest radiography: automated reading vs. interpretation by clinical officers," *International Journal of Tuberculosis and Lung Disease*, vol. 17, no. 12, pp. 1613–1620, 2013.
9. S. Rajaraman, K. Silamut, M. Hossain, I. Ersoy, R. Maude, S. Jaeger, G. Thoma, and S. Antani, "Understanding the learned behavior of customized convolutional neural networks toward malaria parasite detection in thin blood smear images," *Journal of Medical Imaging*, vol. 5, no. 3, pp. 34501–34511, 2018.
10. S. Rajaraman, S. Antani, M. Poostchi, K. Silamut, M. Hossain, R. Maude, S. Jaeger, and G. Thoma, "Pre-trained convolutional neural networks as feature extractors toward improved malaria parasite detection in thin blood smear images," *PeerJ*, vol. 6, p. e4568, 2018.
11. R. Sivaramakrishnan, S. Antani, and S. Jaeger, "Visualizing deep learning activations for improved malaria cell classification," in First Workshop Medical Informatics and Healthcare (MIH 2017), held in Halifax, Nova Scotia, Canada, pp. 40–47, 2017.
12. A. Karargyris, J. Siegelman, D. Tzortzis, S. Jaeger, S. Candemir, Z. Xue, K. C. Santosh, S. Vajda, S. Antani, L. Folio, and G. Thoma, "Combination of texture and shape features to detect pulmonary abnormalities in digital chest X-rays," *International Journal of Computer Assisted Radiology and Surgery*, 2016.
13. K. C. Santosh, S. Vajda, S. Antani, and G. Thoma, "Edge map analysis in chest X-rays for automatic pulmonary abnormality screening," *International Journal of Computer Assisted Radiology and Surgery*, vol. 11, no. 9, pp. 1637–1646, 2016.
14. K. C. Santosh and S. Antani, "Automated chest x-ray screening: can lung region symmetry help detect pulmonary abnormalities?" IEEE Transactions on Medical Imaging, 2018.
15. S. Vajda, A. Karargyris, S. Jaeger, K. C. Santosh, S. Candemir, Z. Xue, S. Antani, and G. Thoma, "Feature selection for automatic tuberculosis screening in frontal chest radiographs," *Journal of Medical Systems*, 2018.
16. F. Ahmad Khan, T. Pande, B. Tessema, R. Song, A. Benedetti, M. Pai, K. Lönnroth, and C. M. Denkinger, "Computer-aided reading of tuberculosis chest radiography: moving the research agenda forward to inform policy," *European Respiratory Journal*, 2017.
17. R. M. Haralick, K. Shanmugam, and I. Dinstein, "Textural features for image classification," IEEE Transactions on Systems, Man, and Cybernetics, vol. SMC-3, no. 6, pp. 610–621, 1973.

18. B. Van Ginneken, S. Katsuragawa, B. M. Ter Haar Romeny, K. Doi, and M. A. Viergever, "Automatic detection of abnormalities in chest radiographs using local texture analysis," IEEE Transactions on Medical Imaging, vol. 21, no. 2, pp. 139–149, 2002.
19. L. Hogeweg, C. Mol, P. A. De Jong, R. Dawson, H. Ayles, and B. Van Ginneken, "Fusion of local and global detection systems to detect tuberculosis in chest radiographs," *Lecture Notes in Computer Science* (including subseries Lecture *Notes in Artificial Intelligence and Lecture Notes in Bioinformatics)*, vol. 6363 LNCS, no. PART 3, pp. 650–657, 2010.
20. S. Jaeger, S. Candemir, S. Antani, Y. J. Wang, P.X. Lu, and G. Thoma, "Two public chest X-ray datasets for computer-aided screening of pulmonary diseases.," *Quantitative Imaging in Medicine and Surgery*, vol. 4, no. 6, pp. 475–477, 2014.
21. A. Chauhan, D. Chauhan, and C. Rout, "Role of gist and PHOG features in computer-aided diagnosis of tuberculosis without segmentation," *PLoS One*, vol. 9, no. 11, pp. 1–12, 2014.
22. J. Melendez, C. I. Sánchez, R. H. H. M. Philipsen, P. Maduskar, R. Dawson, G. Theron, K. Dheda, and B. V. Ginneken, "An automated tuberculosis screening strategy combining X-ray-based computer-aided detection and clinical information," *Scientific Reports*, vol. 6, p. 25265, 2016.
23. T. Pande, C. Cohen, M. Pai, and F. Ahmad Khan, "Computer-aided detection of pulmonary tuberculosis on digital chest radiographs: a systematic review," *International Journal of Tuberculosis and Lung Disease*, vol. 20, no. 9, 2016.
24. S. Candemir, S. Jaeger, K. Palaniappan, J. P. Musco, R. K. Singh, Z. Xue, A. Karargyris, S. Antani, G. Thoma, and C. J. McDonald, "Lung segmentation in chest radiographs using anatomical atlases with nonrigid registration," IEEE Transactions on Medical Imaging, vol. 33, no. 2, pp. 577–590, 2014.
25. J. H. Tan, U. R. Acharya, C. Tan, K. T. Abraham, and C. M. Lim, "Computer-assisted diagnosis of tuberculosis: a first order statistical approach to chest radiograph," *Journal of Medical Systems*, vol. 36, no. 5, pp. 2751–2759, 2012.
26. U. Avni, H. Greenspan, E. Konen, M. Sharon, and J. Goldberger, "X-ray categorization and retrieval on the organ and pathology level, using patch-based visual words," IEEE Transactions on Medical Imaging, vol. 30, no. 3, pp. 733–746, 2011.
27. J. Schmidhuber, "Deep learning in neural networks: an overview," *Neural Networks*, vol. 61, pp. 85–117, 2015.
28. Y. LeCun, B. Yoshua, and H. Geoffrey, "Deep learning," *Nature*, vol. 521, no. 7553, pp. 436–444, 2015.
29. N. Srivastava, G. Hinton, A. Krizhevsky, I. Sutskever, and R. Salakhutdinov, "Dropout: a simple way to prevent neural networks from overfitting," *Journal of Machine Learning Research*, vol. 15, pp. 1929–1958, 2014.
30. A. Ben Abacha, S. Gayen, J. J. Lau, S. Rajaraman, and D. Demner-Fushman, "NLM at ImageCLEF 2018 Visual Question Answering in the medical domain," held in Avignon, France, in *CEUR Workshop Proceedings*, 2018.
31. S. Rajaraman, S. Antani, Z. Xue, S. Candemir, and S. Jaeger, "Visualizing abnormalities in chest radiographs through salient network activations in deep learning," in *Life Sciences Conference (LSC), 2017 IEEE*, held in Sydney, NSW, Australia, pp. 71–74.
32. J. Deng, Wei Dong, R. Socher, Li-Jia Li, Kai Li, and Li Fei-Fei, "ImageNet: a large-scale hierarchical image database," in *2009 IEEE Conference on Computer Vision and Pattern Recognition*, held in Miami, Florida, pp. 248–255.

33. A. S. Razavian, H. Azizpour, J. Sullivan, and S. Carlsson, "CNN features off-the-shelf: an astounding baseline for recognition," in *IEEE Computer Society Conference on Computer Vision and Pattern Recognition Workshops*, held in Columbus, Ohio, pp. 512–519.
34. A. Krizhevsky, I. Sutskever and G. E. Hinton, "ImageNet classification with deep convolutional neural networks," *Communications of the ACM*, vol. 60, no. 6, pp. 84–90, 2017.
35. K. Simonyan and A. Zisserman, "Very deep convolutional networks for large-scale image recognition," *International Conference on Learning Representations*, held in San Diego, California, 2015.
36. C. Szegedy, V. Vanhoucke, S. Ioffe, J. Shlens, and Z. Wojna, "Rethinking the inception architecture for computer vision," in *Proceedings of the IEEE Conference on Computer Vision and Pattern Recognition*, held in Las Vegas, Nevada, pp. 2818–2826, 2016.
37. K. He, X. Zhang, S. Ren, and J. Sun, "Deep residual learning for image recognition," *arXiv Prepr. arXiv1512.03385v1*, vol. 7, no. 3, pp. 171–180, 2015.
38. F. Bousetouane and B. Morris, "Off-the-shelf CNN features for fine-grained classification of vessels in a maritime environment," *Lecture Notes in Computer Science* (including subseries Lecture *Notes in Artificial Intelligence and Lecture Notes in Bioinformatics)*, vol. 9475, pp. 379–388, 2015.
39. Y. Bar, I. Diamant, L. Wolf, S. Lieberman, E. Konen, and H. Greenspan, "Chest pathology detection using deep learning with non-medical training," *Proceedings – International Symposium on Biomedical Imaging*, held in Brooklyn Bridge, New York, vol. 2015–July, pp. 294–297, 2015.
40. T. Ojala, M. Pietikäinen, and T. Mäenpää, "Multiresolution gray-scale and rotation invariant texture classification with local binary patterns," *IEEE Transactions on Pattern Analysis and Machine Intelligence*, vol. 24, no. 7, pp. 971–987, 2002.
41. A. Oliva and A. Torralba, "Modeling the shape of the scene: a holistic representation of the spatial envelope," *International Journal of Computer Vision*, vol. 42, no. 3, pp. 145–175, 2001.
42. A. Bergamo and L. Torresani, "PiCoDes: learning a compact code for novel-category recognition," *Advances in Neural Information Processing Systems*, vol. 24, pp. 2088–2096, 2011.
43. S. Hwang, H.E. Kim, J. Jeong, and H.J. Kim, "A novel approach for tuberculosis screening based on deep convolutional neural networks," in *SPIE Medical Imaging*, held in San Diego, California, 2016, p. 97852W.
44. R. Singh, M. K. Kalra, C. Nitiwarangkul, J. Patti, F. Homayounieh, A. Padole, P. Rao, P. Putha, V. V. Muse, A. Sharma, and S. R. Digumarthy, "Deep learning in chest radiography: detection of findings and presence of change," *PLoS One*, vol. 13, no. 10, p. e0204155, 2018.
45. U. K. Lopes and J. F. Valiati, "Pre-trained convolutional neural networks as feature extractors for tuberculosis detection," *Computers in Biology and Medicine*, vol. 89, pp. 135–143, 2017.
46. L. K. Hansen and P. Salamon, "Neural network ensembles," *IEEE Transactions on Pattern Analysis and Machine Intelligence*, vol. 12, no. 10, pp. 993–1001, 1990.
47. L. Breiman, "Bagging predictors," *Machine Learning*, vol. 24, no. 2, pp. 123–140, 1996.
48. Y. Freund and R. E. Schapire, "A decision-theoretic generalization of on-line learning and an application to boosting," *Journal of Computer and System Sciences*, vol. 55, no. 1, pp. 119–139, 1997.

49. L. Lam and S. Y. Suen, "Application of majority voting to pattern recognition: an analysis of its behavior and performance," *IEEE Transactions on Systems, Man, and Cybernetics – Part A*, vol. 27, no. 5, pp. 553–568, 1997.
50. G. Fumera and F. Roli, "A theoretical and experimental analysis of linear combiners for multiple classifier systems," *IEEE Transactions on Pattern Analysis and Machine Intelligence*, vol. 27, no. 6, pp. 942–956, 2005.
51. D. H. Wolpert, "Stacked generalization," *Neural Networks*, vol. 5, no. 2, pp. 241–259, 1992.
52. P. Lakhani and B. Sundaram, "Deep learning at chest radiography: automated classification of pulmonary tuberculosis by using convolutional neural networks," *Radiology*, vol. 000, no. 0, p. 162326, 2017.
53. C. Liu, J. Yuen, A. Torralba, J. Sivic, and W. T. Freeman, "SIFT flow: dense correspondence across different scenes," *Lecture Notes in Computer Science* (including subseries Lecture *Notes in Artificial Intelligence and Lecture Notes in Bioinformatics)*, vol. 5304 LNCS, no. PART 3, pp. 28–42, 2008.
54. Y. Boykov and G. Funka-Lea, "Graph cuts and efficient N-D image segmentation," *International Journal of Computer Vision*, vol. 70, no. 2, pp. 109–131, 2006.
55. S. E. Grigorescu, N. Petkov, and P. Kruizinga, "Comparison of texture features based on Gabor filters.," *IEEE Trans. Image Process.*, vol. 11, no. 10, pp. 1160–1167, 2002.
56. N. Dalal and W. Triggs, "Histograms of oriented gradients for human detection," 2005 IEEE Computer Society Conference on Computer Vision and Pattern Recognition *CVPR05*, held in San Diego, California, vol. 1, no. 3, pp. 886–893, 2004.
57. H. Bay, A. Ess, T. Tuytelaars, and L. Van Gool, "Speeded-up robust features (SURF)," *Computer Vision and Image Understanding*, vol. 110, no. 3, pp. 346–359, 2008.
58. J. Mockus, "On Bayesian methods for seeking the extremum," *IFIP Tech. Conf. Optimization Techniques*, pp. 400–404, 1974.
59. S. Ioffe and C. Szegedy, "Batch normalization: accelerating deep network training by reducing internal covariate shift," in Proceedings of the 32nd International Conference on Machine Learning, held in Lille, France, pp. 448–456, 2015.
60. K. Ashraf, B. Wu, F. N. Iandola, M. W. Moskewicz, and K. Keutzer, "Shallow networks for high-accuracy road object-detection," in Proceedings of the 3rd International Conference on Vehicle Technology and Intelligent Transport Systems, VEHITS 2017, Porto, Portugal, April 22-24, pp. 33–40, 2017.
61. T. G. Dietterich, "Ensemble methods in machine learning," *Multiple Classifier Systems*, vol. 1857, pp. 1–15, 2000.
62. A. Esteva, B. Kuprel, R. A. Novoa, J. Ko, S. M. Swetter, H. M. Blau, and S. Thrun, "Dermatologist-level classification of skin cancer with deep neural networks," *Nature*, vol. 542, no. 7639, pp. 115–118, 2017.

2

The Role of Artificial Intelligence (AI) in Medical Imaging: General Radiologic and Urologic Applications

Diboro Kanabolo and Mohan S. Gundeti

CONTENTS

2.1 Introduction to Artificial Intelligence (AI)

Artificial Intelligence (AI) permeates many different sectors of industry, including financial/banking, commerce, social media, and health. Several domains of technology are in active development. These include *audio processing*: speech recognition, music/voice identification; *computer vision*: facial or object recognition; *graph analytics*: as in film recommendations, or mapping directions; *language processing*: for example, machine translation, and machine-based question/answer; as well as *time series*, as in stock forecasting. More sophisticated software programs may use any combination of these domains. A popular example to date is autonomous driving, which may use graph analytics, computer vision, and audio processing [1].

2.1.1 Terminology

The vocabulary below will aid in the understanding of details discussed in the discussion to follow. Of note, this brief glossary will only continue to enlarge with the expansion of AI technology.

In this section, we will define the following terms in alphabetical order: artificial intelligence, computer-aided detection, computer-aided diagnosis, computer-aided triage, classification, deep learning, detection, machine learning, model, neural network, representation learning, segmentation, supervised learning, testing, training, transfer learning, unsupervised learning, and validation.

Artificial intelligence: defined in Merriam Webster's dictionary as the branch of computer science that allows a machine the capability to imitate intelligent human behavior [2]. Two types of AI exist: the first is artificial general intelligence, in which a computer may hypothetically mimic the day-to-day actions of human beings. The second, more appropriate definition to be used in the ensuing discussion is artificial narrow intelligence, in which a computer engages in particular tasks.

- Computer-aided detection (CADe): computer recognition of areas of concern necessitating further evaluation. It does not provide diagnosis [3].
- Computer-aided diagnosis (CADx) is simply a computer's ability to state multiple diagnoses (a differential) or a singular diagnosis, to be followed up by a provider [3].
- Computer-aided triage (CAT) occurs when a computer studies an image and further prioritizes it for review by radiologist, or simply provides a diagnosis, with or without follow up. Triage systems may be utilized commonly in screening protocols, in which the intention is to efficiently reduce the clinical care provided by clinicians [4].
- Classification utilizes the clustering of data points that are of similar properties.
- Deep learning utilizes the concept of deep neural networks, large logic networks organized into three basic layers: input, hidden, and output layers [5, 6]. The input layer processes large amounts of relevant data. The hidden layer tests and compares new data against pre-existing data, classifying and re-classifying in real time, with particular connections weighted with degrees of influence. The output layer utilizes confidence intervals to determine the best outcome from various predicted outcomes [6].
- Detection involves the act of identifying and localizing a finding in an image [7].

- Machine learning: in general, the term refers to the process of teaching a computer to learn from input data without specific programming. As it applies to *radiomics,* the term is used to describe high throughput extraction of quantitative imaging features with the intent of creating minable databases from radiological images [8].
- Modeling is defined as the structural state of a neural logic network, allowing for the transformation of data input value into output [3].
- Neural network is a model or logic network composed of layers with transmission of sequential data inputs consisting of nodes analogous to those of biological neural networks. Representation learning is the subtype of machine learning in which an algorithm may acquire those features that enable it to classify data [3].
- Segmentation is the process of delineating the boundary of a particular finding within an image.
- Supervised learning involves inferring a function from inputs and outputs (training data). This allows for output prediction after any novel input [3].
- Testing involves evaluating the performance of a neural logic network [7].
- Training is the process of selecting the ideal parameters of a neural logic network after sequential, repeated adjustments [7].
- Transfer learning may occur when limited data to solve a novel problem are available, but existing data for a problem of close relevance are plentiful [1].
- Unsupervised learning occurs when outputs are inferred in the absence of input data.
- Validation involves the use of a data subset that is distinct from a training set to adjust parameters of the model [7].

2.1.2 Practical Costs

According to an analysis by the accounting firm Price Waterhouse Coopers (PWC), AI is projected to add $15.7 trillion to global gross domestic product (GDP) by 2030 [9]. The true costs of continued development, secondary to the demand of these potential future economic interests, will only spurn increased investments, with the bulk of such investment grounded in the private sector. As of 2016, the United States federal government annually budgeted $1.1 billion in non-classified AI technology [10]. This is contrasted with the $46 billion investment that the top five original equipment manufacturers in the automotive industry spent in 2015 alone [11].

2.2 Artificial Intelligence in Medicine

A large amount of knowledge must be acquired and tailored to solve complex clinical issues. The early beginnings of AI date to the early 1970s with the explosion of biomedical applications, catalyzed by various seminal conferences around the country. Today, there is potential in medicine to harness "big data," utilizing the large quantities of information generated daily in various settings. Clinical practice is now shifting from episodic analysis of disparate datasets to algorithms relying on consistently updated datasets for improved prediction of patient outcomes [12]. The scope and depth of applications is vast. Machine learning algorithms have already been employed to predict the risk for cardiac arrest in infants, and computer visualization has been utilized for various applications, from cancer detection in radiology to detecting health indicators of mental fatigue [13].

A vast amount of further investment is needed for AI in the application of medicine broadly, and medical imaging technology specifically. These investments include: training datasets with representative images for the purposes of validation for computerized deep neural networks (with periodic updating as necessary); they also include the interoperability framework necessary to uphold these costs—software algorithms for protocoling the analysis and extraction of data with the aid of specialized vendors (in addition to original digital imaging software companies). Transparent stakeholder collaboration is necessary to ensure unified file format, data representation and database architecture.

In a 1970 manuscript, the prominent Professors in Radiology Kurt Rossmann and Bruce Wiley stated that the "central problem in the study of radiographic image quality is to gain knowledge regarding the effect of physical image quality on diagnosis and not necessarily to design 'high-fidelity' systems" [14]. This is a reasonable goal, as costs of development should simply be the minimal sufficient to diagnose lesions of interest, and this has historically happened with the aid of human interpretation. However, the direct costs of this benefit appear to have the capability of meeting Rossmann and Wiley's requirements. Currently, a hospital may install $1,000 graphics processors in its imaging machines in order to increase capacity up to 260 million images per day [15]. These fixed costs are attractive for the consumer (institutions providing medical care) and may indeed be utilized with increasing frequency in the coming years. The institutions are incentivized to expand their investment in technology, given the estimation that up to 10% of costs in US healthcare spending are attributed to radiological imaging. Other stakeholders, including patients, clinicians, regulatory bodies, and hospital administrations may benefit from education on risks, benefits, and limitations of the technology [7]. For these reasons, the true shared costs of development with respect to medical imaging are elusive.

2.3 Artificial Intelligence in Radiology

The task of the radiologist is plural in nature. Radiologists must be able to recognize and interpret patterns in medical images, as well as to consult with physicians from various fields to direct patients' care. Narrow AI, which enables CADe of disease, has been in continual development for decades. Radiologic imaging tasks, for the purposes of CADe, and CADx disease discovery have undergone rapid development in recent years [3]. The field of radiomics is built on the premise of converting images into data from which useful details may be extracted, for the ultimate purpose of providing medical utility. We aim to provide an overview of the various processes and subfields, relating the roles of artificial intelligence in medical imaging.

For the practicing radiologist, optimal interpretation of disease requires optimal image quality. Image interpretation, especially in the context of subtle disease states, can be limited by various factors, both extrinsic and intrinsic to the radiologist. Extrinsic factors may include: the knowledge base of the radiologist, the clinical history of the patient, and his or her detection/characterization thresholds. Intrinsic factors to image quality are based on target object attributes, specifically the geometry and contrast of the targets, and their background/visibility (i.e. gray-scale appearance and detail [local noise and resolution of the image]) [16].

2.3.1 Extrinsic Factors to Image Interpretation

Diagnostic error accounts for approximately 10% of patient deaths, and between 6 and 17% of adverse events occurring during hospitalization. At a total of ~20 million radiology errors per year, and 30,000 practicing radiologists, this is an average of under 700 errors per practicing radiologist [17]. Errors in diagnosis have been associated with clinical reasoning, including: intelligence, knowledge, age, visual psychiatric affect, physical state (fatigue), clinical history of the patient, and gender (male predilection for risk taking). These factors, and the limited access to radiologic specialists for up to 2/3 of the world, encourage a more urgent role for the use of AI in medical imaging, a huge focus of which is machine learning [17].

2.3.2 Intrinsic Factors to Image Quality

2.3.2.1 Geometry

Intrinsic object attributes are arranged in three general geometries: (1) point, (2) line, and (3) extended targets. Point targets (Figure 2.1) are small, with maximum dimensions typically lower than 1 mm. These may include microcalcifications, calculi, or osteophytes. Line targets may have a variable length depending on clinical context. Examples include spicules, septate lines, and

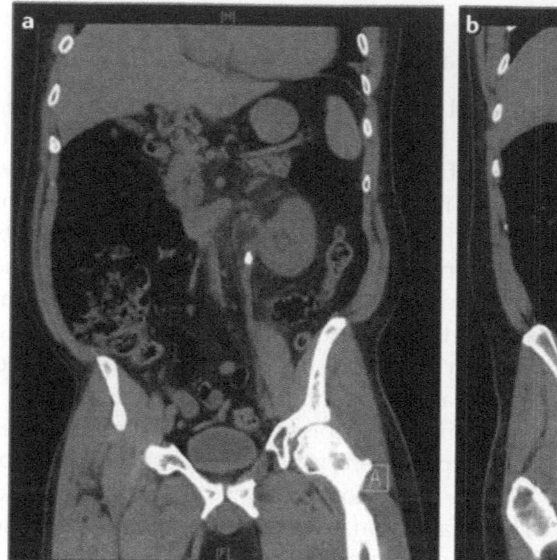

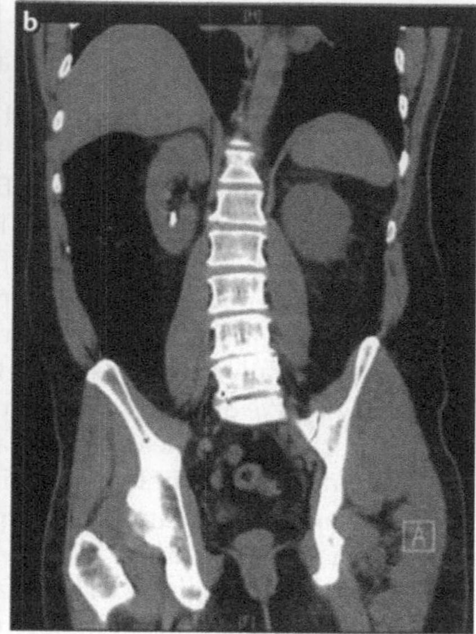

FIGURE 2.1
A coronal demonstration of bilateral 8 mm nephrolithiasis on noncontrast CT, showing point microcalcification. (Reprinted with permission from "An overview of kidney stone imaging techniques" by Brisbane, Bailey, and Sorensen, 2016. *Nature Reviews Urology*, 13, 654–662. 2016.)

lines delineating cortical vs. trabecular bone (Figure 2.2). Extended objects may include tumors, abscesses, and infiltrates [16].

2.3.2.2 Contrast

While in reality, contrast range is on a continuum, we denote here a dichotomy of high and low contrast imaging as per Vyborny, 1997. Examples of high object contrast (Figure 2.3) within the point, line and extended images include a dense microcalcification, tangential pleural calcifications, and calcified granulomas. Fainter microcalcifications, early spicules, and gallstones comprise inherently low contrast point, line, and extended objects, respectively [16]. Note that high object contrast items do not necessarily indicate artificial contrast dye enhancement, though this is a technicality that may vary with the function of other clinical information (renal and liver function, previous allergic responses).

2.3.2.3 Background

The premise of background implies that the radiologic "canvas" on which a particular target finding appears influences its perceptibility. This background may in turn be influenced by not only the gray-scale for which the lesion is displayed, but also the detail of the image itself.

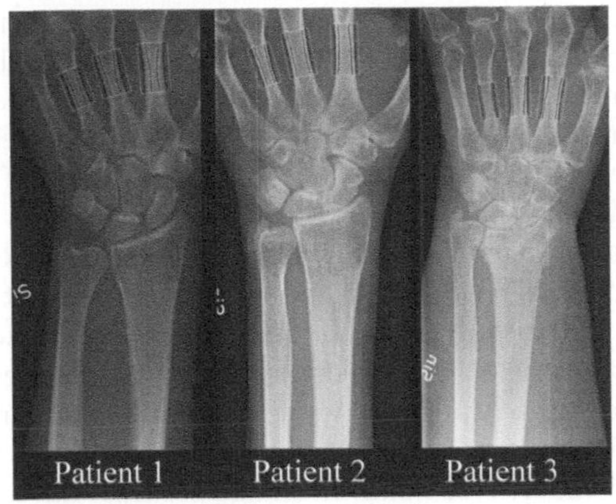

FIGURE 2.2
Delineation of cortical bone thickness (linear). Wrist radiographs used for Dual X-Ray analysis of multiple patients illustrating the cortical bone delineation lines of three patients. (Reprinted with permission from "Digital X-ray radiogrammetry of hand or wrist radiographs can predict hip fracture risk—a study in 5,420 women and 2,837 men" by Wilczek, Kälvesten, Algulin, Beiki, and Brismar. *European Radiology*, 23, 1383–1391. 2012.)

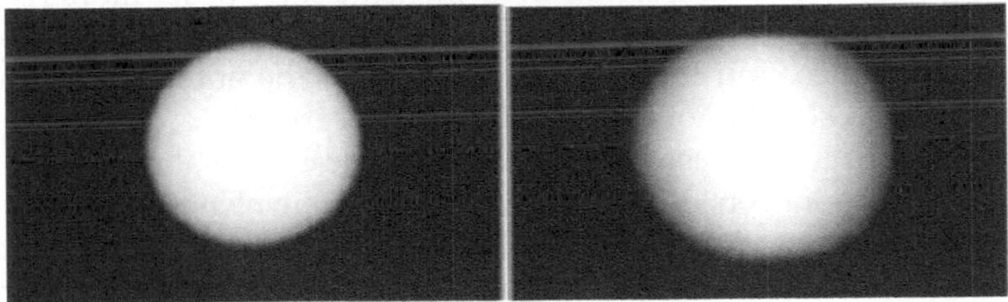

FIGURE 2.3
High vs. low object contrast for circular object. (Cropped and reprinted with permission. "Motion-blur-free video shooting system based on frame-by-frame intermittent tracking" by Inoue, Jiang, Matsumoto, Takaki, and Ishii is licensed under CC BY 4.0.)

Gray-scale components depend on (1) dimensions and anatomical properties, (2) amount of radiation, magnetic resonance, or sound wave frequency to which an area is exposed, and (3) sensitometric characteristics of the recording system [16]. These all determine the relative optical densities of the structures under examination. As stated by Doi et al., 1977, radiomics is challenging in part because of the need to match radiation/magnetic resonance exposure and recording system sensitometric characteristics to the anatomy and targets of interest [18]. When adding the complexity of anatomic structures in immediate proximity, different levels of attenuation may be seen based on the specific qualities of the medium through which the sound waves, radiation, or magnetic resonance are travelling.

Intuitively, details from the perspective of an observer increase in visibility with proximity to an image. As this distance becomes infinitesimally smaller, the effects of local noise become evident. During analysis of images in clinical practice, it is often the case that noise seen by this blurring is essential to the recognition of a pathological phenomenon. These might include pulmonary edema seen on chest X-ray, or invasion of cerebral ventricle by tumor. Thus, these effects are conscious to the observer. The inherent lack of detail caused by sub-optimal resolution or unsharpness (local noise) of a target object on an image pose two problems: (1) The size of the target may be overestimated. (2) The contrast enhancement seen between the image and background will be sub-optimal. This is especially true for point target lesions. Lack of sharpness of an object can lead to an overestimation in size secondary to line spread function, which causes broadening due to lack of inherent contrast between an object and its background. The quantitative measures of the influence of detail in image contrast have been extensively studied. The effect of local noise on line border overestimation is quantified as the line spread function and has been assessed in angiograms (line targets of high contrast) [19]. The possibility of minimal contrast enhancement between background and image due to lack of delineation may be a point of considerable tragedy. In clinical practice, this may perhaps be true of point lesions in particular. For example, in routine chest X-rays, the possibility of missing a diagnosis of lung cancer in its inchoate stage poses more risk than overestimating a tumor's size on an initial interpretation. Detail is of the utmost importance, and it is for this reason that the automation of image analysis with AI should not exclude human interpretation.

With the intrinsic characteristics of imaging described above, we see that a significant proportion of the technical factors involved in proper automated interpretation of an image can be overcome with proper allocation of resources to their development. For example, research funds for the role of machine learning in point and line spread function analysis for historically low-resolution imaging, such as radiographs or computed tomography.

Still, for common lesions, it appears that CADe may be of benefit. Independent studies suggest that women receiving regularly scheduled mammograms over a period of 10 years have a 50 to 63% chance of receiving a false positive diagnosis. It is worth noting that in up to 33% of occurrences, two or more radiologists inspecting the same image may disagree on the findings seen on mammogram. However, with the same empirical methods, visual pattern recognition software is at least 5–10% more accurate than physicians, further supporting the roles of experience and bias in radiologists' reporting [20].

Further techniques for CADe have been studied extensively, including image preprocessing techniques to enhance the image quality followed by an adaptive segmentation. Other techniques utilized include voting-based combination utilizing different classifiers: Bayesian network, multilayer perception neural networks, and random forest for classification of lung region symmetry. Other developments of interest include: angular relational signature-based chest radiograph for classification of frontal and lateral X-rays, edge map analysis for abnormal tissue, generalized line histogram technique for rotational artifact detection, and cross-correlation of circle-like elements using a few normalized templates and unsupervised clustering techniques [21–27].

2.3.3 Specific Technical Example of AI in Medical Imaging

The Kohonen self-organized map (KSOM) may be helpful in the understanding of AI's capacity and potential. It is ideal for utilizing artificial bias and sensory experience to enhance accuracy of CADe and CADx when applied to medical imaging. Briefly, it is an artificial neural network (ANN) developed to decrease complexity by representing multidimensional vectors into as little as one dimension. Yet, data is stored in such a way as to maintain topological relationships [28]. This vector quantization may be illustrated with computerized color discrimination in which individual colors are visualized on a spectrum (Figure 2.4, Figure 2.5a). The KSOM is a form of unsupervised feature extraction and classification. Unsupervised feature extraction utilizes images and clinical narrative texts to allow for high throughput application in the analysis of clinical data. The steps in this process include disease detection, lesion segmentation, diagnosis, treatment selection, response assessment via repeat imaging, and using data about a patient to create clinical predictions in regards to potential patient outcomes [29]. In contrast, supervised techniques typically consist of vector pairs and

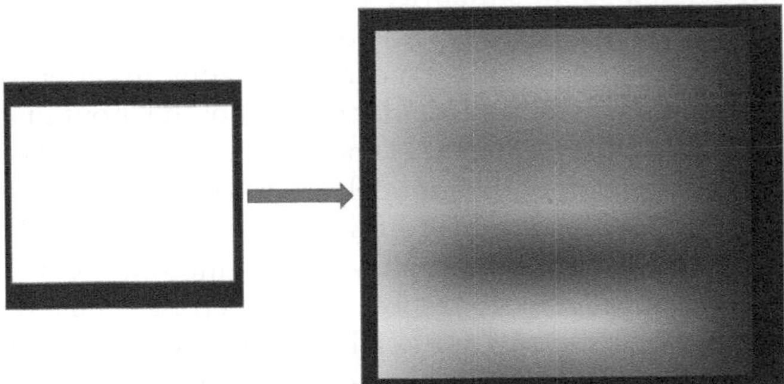

FIGURE 2.4
Color discrimination analogy for illustrating KSOM. The white color spectrum is broken down into its varying wavelengths in a topographically intact manner. It is important to note that regions of the spectrum containing similar properties (wavelengths) are clustered together. Each distinct coordinate in the map has its own component red, green, and blue vectors. When vectors (colors) are randomly selected from a training dataset, all distinct coordinate nodes are examined to assess that which is the most similar to the input vector from the training set. This coordinate will be known as the BMU. A radius for which this best value unit applies is calculated and is known as the BMU neighborhood. The neighboring coordinates are then adjusted (weighted with respect to the BMU), and the process of input vector presentation is repeated. This mechanism is an ideal illustration for understanding unsupervised learning in AI.

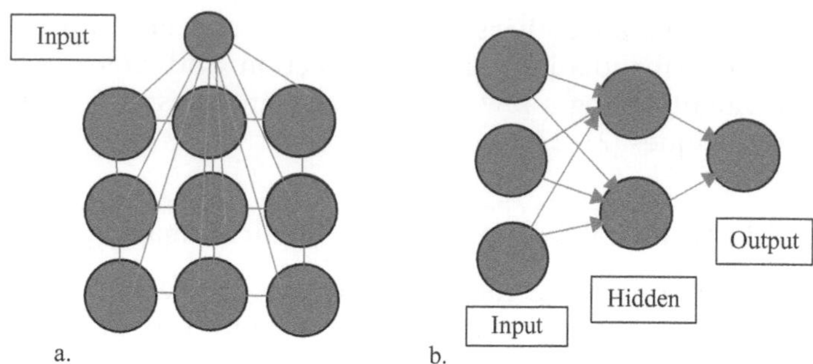

FIGURE 2.5
(a) Kohonen network. Each node of the 3 x 3 lattice represents a vector with a distinct coordinate and contains a particular corresponding weight with respect to the input vectors. (b) An ANN. Each node represents a neuron, and arrows represent specific synapses between them. The input layer feeds are processed by a hidden layer of functions before being expelled into a known result, or output.

include input and output (goal) vectors. When an input is presented to the neural network, the output of the network is compared with this goal vector. When the two contrast, the weights of the networks neurons are adjusted to reduce the error in this output. This process may be seen in neural networking systems with the sole purpose of validating the neural network developed by imaging software of interest. This form of learning, the self-organizing map, can be applied broadly to other neuronal networks.

The KSOM bears a slight similarity to artificial neural networks, in that rather than being a logic network, it is one of a topographical relationship (Figure 2.5b). Training each node is a complex matter. In order for the node to recognize its similarity to a particular element of the input image, it must calculate the weights of its inherent elements. The node with the most similarity to a distinct portion of the input element is known as the best matching unit, or BMU. Counterintuitively, this seems to contradict the necessity of imaging, as each lattice of nodes may contain multiple coordinates with similarly containing weights. To resolve this concern, we must note that the entire lattice may be considered as a neighborhood of elements. At each iteration, the radius of the BMU within the neighborhood—otherwise described as the area within the lattice in which BMU is most similar to the weighted input element—is determined. The weights are then adjusted and the radius of the nodes continues to shorten as more BMUs are described with time.

With an appreciation for this complexity, we must note that the conduction of peer review reads, as in radiology-aided clinical interpretation with hope of CADx, is in its infancy. Before this process is popularized, it is a popular conception that the technology be used for the purposes of screening at the population level. For example, detection of diabetes-induced macular edema is currently not financially practical, as 20 million diabetics in the country would need to be screened for a yield of approximately 5%.

2.4 Urologic Applications

As alluded to previously, the automobile has undergone extreme enhancement. From initial prototypes in the early 20th century to the present, with autonomous vehicle development, we have seen significant evolution. Some posit that the development of the da Vinci robotic surgical system developed by Intuitive Surgical Sunnyvale CA (USA) will be no different. Used throughout a variety of urologic procedures, the da Vinci robot is highly valued in surgeries for deep and narrow fields, and when micro-suturing/fine dissection are needed [30]. Automated surgical development with the robot may one day be of immense interest. The difference to be overcome between driving and surgery presumably is the emergent non-linear changes that may occur in a surgical suite—for example, hemorrhage, bowel perforation, etc. Still, the technology is under development and remains full of potential. An automated robot known as the Smart Tissue Autonomous Robot (STAR) has already been shown capable of performing surgical procedures using ex-vivo porcine tissue and in living pigs. STAR's accuracy is due in part to integration of near-infrared fluorescent imaging, as well as 3D quantitative plenoptic imaging. The former is currently used for intra-operative detection of sentinel lymph nodes, but has several applications for intra-operative imaging. This is due to the minimal absorption by abundant molecules, and consequent absence of autofluorescence in the near-infrared range [31]. The latter, plenoptic imaging elaborately computes a 3D point for each pixel in an image through a microlens array and image sensors. While these technologies are under development, they bolster the potential of urologic surgery.

AI also has the potential to enhance skills assessment during real-time cases, providing immediate feedback with or without the aid of live human proctors. The technology has already been used to analyze movement trajectory, for eye tracking, and for gaze mapping data. It may prove to have a role in surgeon credentialing, already being able to categorize surgeons on a gradient from novice to expert [32, 33].

Deep learning has been found to be superior to non-deep learning forms of artificial intelligence, including various image recognition technologies [34]. The technology, feedforward probabilistic neural networks, has been used to analyze prognosis of bladder cancer recurrence by histopathology using morphological and textural nuclear features. It shows precision, with accuracy of 72.3% for recurrent tumors, and 71.1% for no recurrence [35]. Neural networks have also been used by many urologists with favorable results in the interpretation of prostatic and renal images [36]. One such study by Tewari et al. utilized an ANN to predict recurrence of prostate cancer following surgery and radiotherapy. The model used conventional parameters to assess 1,400 patients after radical prostatectomy, accurately predicting PSA progression in 76% of cases. As early as 2003, ANN grading and classification of urothelial carcinomas showed promise. Utilizing the Adaptive Stochastic

On-Line method with respect to morphological and textural nuclei, Grade I, II, and III tumors showed a classification accuracy of 90%, 94.9%, and 97.3%, respectively [37].

Research on AI-guided prostate cancer detection published in 2018 is promising as well, suggesting a 92% sensitivity and 82% specificity when using likelihood maps trained by support vector machine learning (a form of supervised learning technology) using a combination of T2 weighted, apparent diffusion coefficients, and diffusion weighted imaging from 14 patients [38]. This compares to 89% sensitivity and 73% specificity for radiologist-detected lesions [30]. This technology may prove of high impact, as the challenge for urologists remains to distinguish benign from malignant lesions without invasive biopsy. In 1995, Moul and colleagues used a KSOM and back propagation programs to stage non-seminomatous testicular germ cell tumor staging, with a sensitivity of 88% and specificity of 96% [39]. There is a paucity of data published in the last two decades prior to the present on further development of AI as it specifically relates to urologic imaging. A significant proportion of AI urological imaging research involves prostate cancer detection or staging. Abbod et al. reported that 60% of published articles found were related to prostate cancer detection, staging, and prognosis. This compares to 32% for bladder tumors.

Ogiela and Tadeusiewicz have used structural pattern recognition to represent upper urinary tract lesions with the aid of nominal features. In the course of analysis, they utilized expansive tree grammar to segment and filter the images, skeletonize, and then transform them into 2D images with diagrams showing the various contours of the straightened organ. In this way, they applied grammar rules to the urograms of the renal pelvis and calyces together with skeletons obtained using a skeletonizing algorithm. This tree grammar algorithm, $G_{edt} = (\Sigma; \Gamma; r; P; Z)$ has definitions for each component. The entire algorithm, G_{edt}, works in stages. The first stage works to define the renal pelvis, followed by larger renal calyces, then smaller calyces, and finally the renal papillae. The short branches of the renal papillae are found to be concave with respect to the smaller calyces after skeletonization. If the papillae are convex/shortened, an abnormality may be present and detected [40].

2.5 Benefits vs. Disadvantages

The benefits of artificial intelligence are numerous. The first of these includes increased productivity, due in part to absence of need for natural breaks in a 24-hour workday, enabling images to be read continuously. This allows results to be returned to patients quickly, and aids medical decision making.

Secondly, the lack of humanistic implications may be a strength as well. The various biases, lack of knowledge, or clerical errors made in the process of observing an image are minimized with computerization. Third, the cost of instituting a new graphics processor or imaging software is fixed. Over time, not only does this benefit allow for saving of human resources, it facilitates a margin of profit for any healthcare administration that only grows with time. Finally, artificial intelligence will continue to progress in its innovative capacities. Because many hospitals apply marketing strategies, AI may enable hospitals to market to their stakeholders, including potential employees and patients [41].

The disadvantages of artificial intelligence must also be considered. First, AI imaging algorithms and software may be able to compete for existing human labor more cost-effectively—it is estimated that AI will eliminate 1.8 million jobs by 2020. This disadvantage is tempered by the increase in employment—in 2017, it was estimated that AI will generate 2.3 million jobs by 2020, and will have a net gain of 2 million jobs by 2025 [42]. Secondly, the personal connection of reviewing an image will continue to be a necessary component of radiology; however, with further development and validation of imaging and interpretation software, there is a reasonable fear concerning the loss of the human review. This is the case of the development of STAR surgery. In the ensuing decades of its development, enhancements of imaging technology may facilitate the automation of surgery. In time, the personal trust a patient places in a surgeon's skills may grow increasingly tenuous.

2.6 Future Considerations

Looking to the future, AI's continued development will likely follow Satya Nadella's three phases for the many technological breakthroughs that have preceded it. The first, invention and design of the technology itself; the second, retrofitting (e.g. engineers receive new training, traditional radiologic equipment is redesigned and rebuilt); the third, navigation of the dissonance, distortion, and dislocation, where challenging novel questions are raised [36]. These may include: what the function of the physician will be when radiomics enables the detection of trends in particular illnesses, or whether CADe or CADx may guide management without the aid of the radiologist.

Many agree that AI should augment, rather than replace human ability. Amongst other considerations previously discussed, this must also be infused with the applicable protections for transparency, as well as privacy and security. This can be accomplished, but it will require a concerted effort amongst all stakeholders, including the requisite regulatory bodies governing the integration and incorporation of these powerful tools into practice.

References

1. S. De Brule, "What is artificial intelligence?" *MachineLearnings*, 2017. Available at: https://machinelearnings.co/how-to-prepare-your-career-for-artificial-intelligence-driven-automation-1bb153759b3b.
2. Merriam-Webster Definition of Artificial Intelligence, Merriam Webster, 2018. Available at: https://www.merriam-webster.com/dictionary/artificial%20intelligence. Accessed May 26, 2018.
3. M. L. Giger, "Machine learning in medical imaging," *Journal of the American College of Radiology*, vol. 15, no. 3, pp. 512–520, 2018.
4. FDA, US, "Guidance for Industry and Food and Drug Administration Staff: Computer-Assisted Detection Devices Applied to Radiology Images and Radiology Device Data—Premarket Notification [510 (k)] Submissions," 2012.
5. S. Nadella, *Hit Refresh: The Quest to Rediscover Microsoft's Soul and Imagine a Better Future for Everyone*, HarperCollins Publishers, New York, 2017.
6. S. Kim, "AI Coming to a Portfolio Near You." Barron's, 2018.
7. A. Tang *et al.* "Canadian Association of Radiologists white paper on artificial intelligence in radiology," *Canadian Association of Radiologists Journal*, 2018.
8. National Academies of Sciences, Engineering, and Medicine, *Improving Diagnosis in Health Care*. National Academies Press, 2016.
9. Pwc, "Sizing the prize: what's the real value of AI for your business and how can you capitalise?" *PwC*. Available at: https://www.pwc.com/gx/en/issues/analytics/assets/pwc-ai-analysis-sizing-the-prize-report.pdf. Accessed May 26, 2018.
10. G. Brockman, "The dawn of artificial intelligence," Testimony before U.S. Senate Subcommittee on Space, Science, and Competitiveness, November 30, 2016.
11. Richard Viereckl *et al.* "Connected car report 2016: opportunities, risk, and turmoil on the road to autonomous vehicles," *PwC*, 2016. Available at: http://www.strategyand.pwc. com/reports/connected-car-2016-study.
12. Z. Obermeyer and E.J. Emanuel, "Predicting the future—big data, machine learning, and clinical medicine," *The New England Journal of Medicine*, vol. 375, no. 13, pp. 1216, 2016.
13. Y. Yamada and M. Kobayashi, "Detecting mental fatigue from eye-tracking data gathered while watching video: evaluation in younger and older adults," *Artificial Intelligence in Medicine*, 2018.
14. K. Rossmann and B. E. Wiley, "The central problem in the study of radiographic image quality," *Radiology*, vol. 96, no. 1, pp. 113–118, 1970.
15. M. Molteni, "If you look at X-Rays or Moles for a Living, AI is coming for your job," Wired, 2017. Available at: https://www.wired.com/2017/01/look-x-rays-moles-living-aicoming-job/
16. C. J. Vyborny, "Image quality and the clinical radiographic examination," *Radiographics*, vol. 17, no. 2, pp. 479–498, 1997.
17. World Health Organization, *Baseline Country Survey on Medical Devices 2010*, WHO Press, Geneva, Switzerland, 2011.
18. K. Doi, K. Rossmann, and A. G. Haus, "Image quality and patient exposure in diagnostic radiology," *Photographic Science and Engineering*, vol. 21, no. 5, pp. 269–277, 1977.

19. K. Rossmann, A. G. Haus, and G. Dobben, "Improvement in the image quality of cerebral angiograms," *Radiology*, vol. 96, no. 2, 361–366, 1970.
20. R. Pearl, Artificial intelligence in healthcare: Separating reality from hype. *Forbes*. 2018.
21. K. C. Santosh and Laurent Wendling, "Angular relational signature-based chest radiograph image view classification," *Medical & Biological Engineering & Computing*, pp. 1–12, 2018.
22. K. C. Santosh and Sameer Antani, "Automated chest x-ray screening: can lung region symmetry help detect pulmonary abnormalities?" *IEEE Transactions on Medical Imaging*, vol. 37, no. 5, pp. 1168–1177, 2018.
23. S. Vajda *et al.* "Feature selection for automatic tuberculosis screening in frontal chest radiographs," *Journal of Medical Systems*, vol. 42, no. 8, p. 146, 2018.
24. F. T. Zohora, Sameer Antani, and K. C. Santosh, "Circle-like foreign element detection in chest x-rays using normalized cross-correlation and unsupervised clustering," *Medical Imaging 2018: Image Processing*, vol. 10574. International Society for Optics and Photonics, 2018.
25. F. Tuz Zohora and K. C. Santosh, "Foreign circular element detection in chest x-rays for effective automated pulmonary abnormality screening," *International Journal of Computer Vision and Image Processing (IJCVIP)*, vol. 7, no. 2, pp. 36–49, 2017.
26. K. C. Santosh *et al.* "Edge map analysis in chest X-rays for automatic pulmonary abnormality screening," *International Journal of Computer Assisted Radiology and Surgery*, vol. 11, no. 9, pp. 1637–1646, 2016.
27. K. C. Santosh *et al.* "Automatically detecting rotation in chest radiographs using principal rib-orientation measure for quality control," *International Journal of Pattern Recognition and Artificial Intelligence*, vol. 29, no. 02, p. 1557001, 2015.
28. T. Kohonen, "Essentials of the self-organizing map," *Neural Networks*, vol. 37, pp. 52–65, 2013.
29. D. Rubin, "Frontiers of AI in medical imaging for clinical decision making." 2017 Human AI Collaboration: A Dynamic Frontier Conference. Accessed Apr 07, 2018.
30. M. Honda *et al.* "Current status of robotic surgery in urology," *Asian Journal of Endoscopic Surgery*, vol. 10, no. 4, pp. 372–381, 2017.
31. R. G. Pleijhuis *et al.* "Near-infrared fluorescence (NIRF) imaging in breast-conserving surgery: assessing intraoperative techniques in tissue-simulating breast phantoms," *European Journal of Surgical Oncology (EJSO)*, vol. 37, no. 1, pp. 32–39, 2011.
32. M. J. Fard *et al.* "Automated robot-assisted surgical skill evaluation: predictive analytics approach," *The International Journal of Medical Robotics and Computer Assisted Surgery*, vol. 14, no. 1, p. e1850, 2018.
33. W. R. Boysen, M. G. Gundeti, "Use of robot-assisted surgery in pediatric urology—have we reached the limits of technology?" *AUANews*, vol. 23, no. 6, pp. 19–20, 2018.
34. Xinggang Wang *et al.* "Searching for prostate cancer by fully automated magnetic resonance imaging classification: deep learning versus non-deep learning," *Scientific Reports*, vol. 7, no. 1, p. 15415, 2017.
35. P. Spyridonos *et al.* "A prognostic-classification system based on a probabilistic NN for predicting urine bladder cancer recurrence," in *Digital Signal Processing, 2002. 2002 14th International Conference on Vol.* 2. IEEE, 2002.

36. M. F. Abbod *et al.* "Application of artificial intelligence to the management of urological cancer," *The Journal of Urology*, vol. 178, no. 4, pp. 1150–1156, 2007.
37. D. K. Tasoulis *et al.* "Urinary bladder tumor grade diagnosis using on-line trained neural networks," in *International Conference on Knowledge-Based and Intelligent Information and Engineering Systems*, 2003, Springer, Berlin, Heidelberg, Germany.
38. Y. Oishi *et al.* "Automated diagnosis of prostate cancer location by artificial intelligence in multiparametric MRI," *European Urology Supplements*, vol. 17, no. 2, e888–e889, 2018.
39. J. W. Moul *et al.* "Neural network analysis of quantitative histological factors to predict pathological stage in clinical stage I nonseminomatous testicular cancer," *The Journal of Urology*, vol. 153, no. 5, pp. 1674–1677, 1995.
40. M. R. Ogiela and Ryszard Tadeusiewicz, "Artificial intelligence structural imaging techniques in visual pattern analysis and medical data understanding," *Pattern Recognition*, vol. 36, no. 10, pp. 2441–2452, 2003.
41. G. A. Okwandu, "Marketing strategies of hospital service organizations in Nigeria: a study of selected privately owned hospitals in Port Harcourt," *Journal of Hospital Marketing & Public Relations*, vol. 14, no. 1, pp. 45–57, 2002.
42. Yen Nee Lee, "Robots 'are here to give us a promotion,' not take away jobs, Gartner says." *CNBC*, 2017. Available at: https://www.cnbc.com/2017/12/18/artificial-intelligence-will-create-more-jobs-than-it-ends-gartner.html?__source=facebook%7Ctech.

3

Early Detection of Epileptic Seizures Based on Scalp EEG Signals

Abhishek Agrawal, Lalit Garg, Eliazar Elisha Audu, Ram Bilas Pachori, and Justin H.G. Dauwels

CONTENTS

3.1 Introduction

Epilepsy is a serious neurological disease that has a substantial impact on the world population. Worldwide, about 50 million people (1% of the world population) are known to have been suffering from epilepsy, and 25% of them have medically resistant forms of epilepsy [2, 52]. Medically resistant epilepsy poses a significant challenge to the clinical management of the disorder and the full range of mechanisms responsible for drug resistance epilepsy is yet to be determined [52, 63]. This makes the problem of epileptic seizure diagnosis, detection, and mitigation extremely important. A study [14] reveals that almost half of the epilepsy patients are children or young adults (aged 0–24). Epilepsy is a medical condition associated with abnormal electrical discharge in the brain. This interferes with the normal central nervous system activities due to seizure episodes, which are recurrent in nature [6, 23, 27]. The level of a patient's consciousness and loss of bodily control may vary depending on the severity of the seizure.

Epileptic seizure can be classified based on clinical semiology, interictal electroencephalography (EEG) recordings, and ictal EEG patterns as defined by the International League Against Epilepsy [51]. Epileptic seizure associated with ictal seizure has a range of manifestation during seizure episodes. During an aura ictal seizure, the subject may experience certain subtle sensations such as tingling, numbness, hallucination, and a smell or odour without any clinical sign being observed [46, 51]. This range of sensorial or experiential symptoms depends on the sphere affected by the epileptic seizure. Thus, ictal semiology includes aura, mutonomic, dialeptic, and motor seizures, beside special seizures [51]. Epileptic seizures can be classified as generalized or localized seizures. In generalized seizures, the episode starts from a source and spreads to other sections of the brain, while localized seizures affect only a section of the brain such as the temporal lobe [29]. A class of generalized seizures called "myoclonic" has a clinical manifestation that is very subtle and often missed by some of the seizures-detection techniques [6, 57]. It manifests as a short abrupt flexion involving only the lower arm or a leg movement. Other seizure classifications are partial seizures (simple and complex seizures), primary generalized seizures, and unclassified seizures [57]. In a simple partial seizure, the subject is not impaired and is conscious, while, in a complex partial seizure, impairment of consciousness may accompany the seizure. Symptoms of motor, somatosensory, special sensory, autonomic, and behavioural signs are observed. Primary generalized seizures include absence, myoclonic, clonic, and tonic seizures [57]. Regardless of the type of seizure, clinical management of drug-resistant epilepsy places a significant strain on finances, and on social and economic wellbeing. This situation has a significant impact on the life of people living with epilepsy. Therefore, the aim of this study is to develop a novel method of feature extraction for effective characterization of epilepsy seizures using a supervised machine learning approach.

The current research in automatic epileptic seizure detection is directed towards the development of a robust algorithm that can satisfy the cost of computation and performance expectations. The two essential components of this algorithm are a feature extraction method and a classifier. A number of different approaches have been used to extract features from the EEG data, such as time and frequency domain analysis [30], empirical mode decomposition (EMD) [21, 58, 61], and spectral power analysis [13]. In recent times, a feature extraction method based on the application of eigenvalue decomposition of Hankel matrix (EVDHM) and its improved version (IEVDHM) have been demonstrated for analysis of non-stationary signals [62, 64]. These data-driven approaches decompose signal in terms of eigenvalues and extract features using the Hilbert transform. Similarly, the problem of cross term associated with the Wigner and Wigner-Ville time-frequency methods can be reduced using the IEVDHM [15, 65]. A novel time-frequency EEG signal analysis for epileptic seizures based on tunable-Q wavelet transform and fractal dimension is shown to be effective to characterize the pattern of epileptic seizures [16]. Signals can be analyzed using bi-orthogonal linear phase wavelet filter banks [15], empirical wavelet transform [66], and key-point based local binary pattern of EEG signal [72].

Classification of data to detect seizure and seizure-free intervals has traditionally been achieved with the help of machine learning–based classifiers such as support vector machines (SVMs) [5, 9, 68]. However, a functional algorithm that is compatible with both computational costs and power constraints for a real-time seizure detector is yet to be realized. To achieve this, several studies [21, 49] have tried techniques such as channel selection and feature ranking, which result in reduction of feature size and computational complexity. In a recent study [5], dynamic time-warping kernel has been proposed to enable SVM to classify variable-length sequences of feature vectors. The study [5] also highlights the importance of a detector that can track the temporal evolution characteristics of seizures and has used dynamic features to obtain encouraging results.

3.2 Electroencephalogram

The commonly used electrophysiologic method of sampling electrical activities or signal from the brain is EEG [10]. It is a brain imaging technique that records signal as a summation total of the post-synaptic potential of neurons generated from the cerebral cortex [10, 43, 70]. Signals propagate through nerve tissues, meninges, and skull to the site where testing electrodes are placed. Extracranial EEG is a noninvasive technique that does not involve incision to position the sensing electrode. Electrodes are spatially distributed over the scalp to capture a portion of electrical activities generated by a neural circuit in the brain. Intracranial EEG (iEEG) imaging techniques require incision to surgically implant microelectrodes into a section or region of the brain to be investigated [24, 55]. The output of an EEG is a graphical display of an amplitude variation over time, which encodes information about the state and pathological conditions of the brain [32, 43, 55, 70]. This method of brain imaging is becoming a gold standard in biomedical data acquisition research and human–machine interactions. A scalp EEG provides ameans of studying the progression of epilepsy or evaluating the performance of an anticonvulsant that is easy to set up and use. In addition, the scalp method has the ability to sample signals with great temporal resolutions and localizes electrical activities with a good topographic spatial resolution [43, 50]. However, it is limited by the presence of artifacts, noise and interference in the EEG signal [34, 45]. The intracranial iEEG method provides a higher resolution of EEG data, usually, at a larger and different scale of activities than that of a scalp EEG [47]. An iEEG measures local field potential (LFP) from the affected section of the brain, which can be used to identify and localize epileptogenic zones for surgery [28]. It is less affected by volume conduction and interference from muscle contraction or movement. Both classes of EEG imaging techniques have been used in the analysis of epilepsy-related seizures with proven and satisfactory results [9, 52]. Several studies indicate that the use of a particular type of EEG depends on the application [43, 70]. In addition, the type of montage used in recording the brain's electrical phenomena can affect the quality of the signal [17, 42]. However, regardless of the type of EEG acquisition being used and the signal representations montage, an EEG is a complex waveform consisting of superimposed signals from different sources, which requires robust signal processing techniques to extract features.

In this study, our primary objective has been to identify and test robust features to reach acceptable values of performance metrics for a real-time seizure detector. These features are extracted from the intrinsic mode function (IMF) components that we obtain after implementing the EMD algorithm on the data. So far, we have tested three features: mean frequency, power, and another feature called "PowF," which was defined using the former two. Specifically, mean frequency has been extensively tested and has produced promising results with detection accuracy (sensitivity) of 100% for 10 patients and 80 % for the four patients for whom substantial data for seizure and seizure-free intervals are available. Our algorithm correctly predicted seizures in the first 10 patients with an average specificity of 95.5%. For the next four patients, it achieved an average specificity of 85.4%. The average latency for the former group was 2.53 seconds and that of the latter group was 3.65 seconds. We tested the entire dataset of the patients available to us. We also obtained mixed results for the remaining 10 patients with the algorithm performing well for most of them on the specificity front but poorly on the sensitivity front. This is possibly due to a data/implementation discrepancy. In general, we observed a trade-off between detection latency (delay in detection from seizure onset) and specificity values while evaluating the performance of classification algorithms.

3.3 EEG Signal Processing

In this section, the data preprocessing and segmentation is presented. The focus here is to construct a patient-specific feature space vector using publicly available EEG data from the CHB-MIT scalp EEG database [66].

3.3.1 EEG Data Preprocessing

An EEG dataset collected at Children's Hospital Boston for the evaluation of epilepsy surgery in pediatric patients CHB-MIT scalp EEG database [66] and made publicly available [49, 68] has been used. This dataset consists of continuous scalp EEG recordings from 24 patients under the age of 18 suffering from intractable seizures. The pediatric patients from whom it is collected did not suffer only from a specific class of seizures, and the dataset consists of focal, lateral, and generalized seizure onsets. This broadens the scope of the detection problem and applicability of the algorithm. The EEG was sampled at 256 Hz and was recorded using a varying number of channels (greater than 22 for most patients) with scalp electrodes placed as per the international 10–20 system. Overall, this 24-patient dataset contains 933 hours of continuously recorded EEG and 193 seizures. Information about seizure onset and seizure duration within a particular hour-long data subset (stored in .edf file format) is available in metadata files.

For each file, all seizures are separated and combined into what looks like a continuous recording of EEG. This allows arbitrary selection of epoch size of a data chunk for feature extraction. Not all the seizure-free EEG dataset is used. Epoch size is shown to have significant effect on the performance and accuracy of feature classification in supervised machine learning.

Figure 3.1 shows the preprocessing of EEG data by separating them into seizures and seizure-free EEG data. A bandpass filter is applied to filter the EEG signal into delta (0–4 Hz), theta (4–8 Hz), alpha (8–15 Hz), and beta (15–30 Hz). So, for each pattern (seizure and seizure-free) data, there are four different sets of data associated with the frequency bands. The selection of the frequency band to be monitored depends on the method of feature extraction and on the band that gives the most discriminative features between seizures and seizures-free episodes. Different methods have been developed to extract features and accurately classify them to detect the pattern of abnormality associated with seizures [3, 8, 17, 20, Bujega et al., 2019]. In seizure onset detection systems, different techniques and algorithmic models have been developed to improve the quality of EEG signal and identify relevant information in the dataset through a process known as feature extraction. These methods of feature extractions are largely based on suppression of artifacts that interfere with signals of interest and elimination of irrelevant background information. An important pattern in EEG can be processed based on the temporal, spectral, and spatial characteristics of the signal [36, 67]. Some of the methods used for filtering, feature extraction, and dimensionality reduction are wavelet [2, 56, 67], cross-correlation, and chaos theory [35]; hardware-based system on chip (SoC) for EEG data processing and feature extractions [54, 69], and high order spectral [53]. Feature selection is essential in the development of automatic medical diagnostic systems [76].

In practical application, EEG is a high-dimensional data, and reducing the dimensional simplifies the training and classification process of supervised

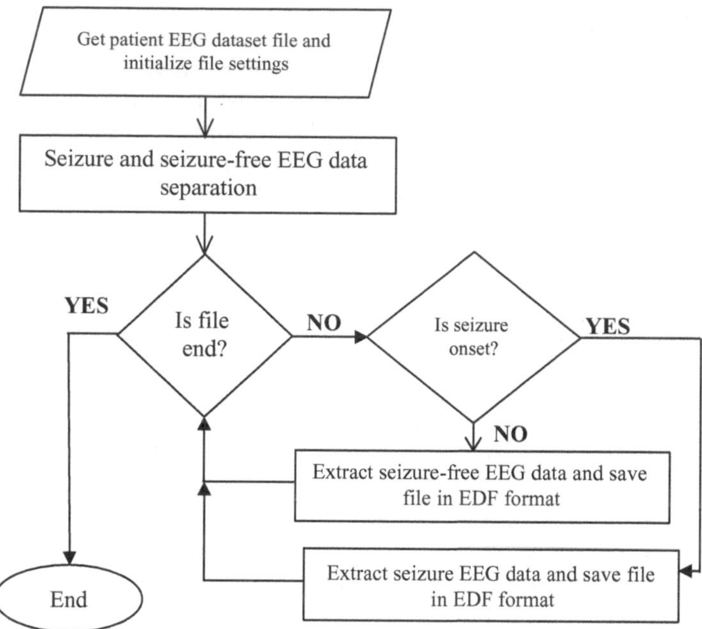

FIGURE 3.1
Preprocessing of scalp EEG data into separate seizure and seizure-free EEG files.

machine learning. Active supervised machine learning with noisy label data and improved traditional classifiers have been demonstrated in pattern classification, which can be used in seizure onset detection systems [18]. As part of implementing an effective and efficient means of compressing information, principal component analysis (PCA) developed in 1933 by Hotelling has been successfully applied. The method acts as an EEG preprocessor to map high-dimensional data into low-dimensional representation by extracting the most important features [36]. It relies on the eigendecomposition method to extract the pattern of similarity from multivariate datasets using the magnitude of the eigenvalues derived from a given dataset [1]. Channels of importance, the principal component, can be identified in EEG using this method and represented as an orthogonal dataset. Thus, significant processing power can be conserved and data redundancy can be reduced [36]. In recent times, an important exploratory statistical signal processing method called common spatial pattern (CSP) has gained significant attention in brain–computer interface. It extracts features using the eigen-decomposition method based on the variance between two given classes of bandpass-filtered EEG signals [26, 31, 44, 45, 69].

In constructing an automatic onset seizure detector, the multidimensional EEG dataset is filtered into frequency bands of interest before features are extracted using a suitable method, as shown in Figure 3.2. The features are arranged into n-dimensional vector space for supervised machine learning classification.

3.3.2 Feature Extraction

Accurate classification of patterns largely relies on the effectiveness of the feature extraction stage. It encodes the characteristics and properties of a dataset. Figure 3.2 shows an automatic feature classification process for early detection of seizures. A statistical method such as CSP has been used to extract features for identification of patterns associated with seizures. CSP is a filtering technique that optimally discriminates between two classes of data based on their variances in such a way that the variance in one class is maximum and minimum for the other class [26, 31, 44, 45, 69]. It can be used to develop a new temporal time series whose variance is optimal for discriminating between features related to a mental task [31, 74]. The filter can functionally reflect the selective activation states of the cortical areas of the brain [31]. The CSP method has shown its effectiveness in extracting topographic patterns of brain rhythmic modulations, which can be used to identify regions of abnormal neurological activities [45, 74]. Another non-statistical method used in EEG feature extraction is the spectral power density of a signal. It shows how a signal evolves or distributes over a frequency scale in time. Several methods based on linear and non-linear time–frequency analysis have been applied to feature extraction [22, 25, 37, 38, 73]. Figure 3.2 shows some of the features used in EEG feature extraction and their classification performance.

Data-driven methods centered on the EMD are shown to very effective in extracting discriminative features. EMD decomposes a signal into secondary components called IMFs for the analysis of non-linear, non-stationary

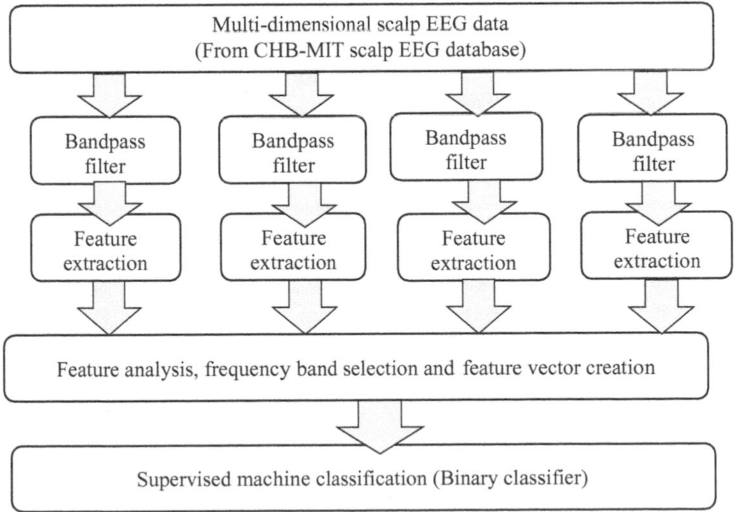

FIGURE 3.2
Automatic feature classification process for early detection of seizures.

signals [11]. In EEG analysis of seizures, decomposing ictal and interictal EEG using EMD into different IMFs results in coefficients of variation and fluctuations [4]. Thus, EMD has shown promising applications in feature extraction for seizures detection [11, 12, 59]. In a recent study [33], dynamic time warping kernel has been proposed to classify variable-length sequences of feature vectors. The study [33] also highlights the importance of a detector that can track the temporal evolution characteristics of seizures and has used dynamic features to obtain encouraging results.

The performance evaluation metrics used in Figure 3.3 is given as [60]:

$$\text{Accuracy}(\%) = \frac{\text{Total number of correctly classified instances}}{\text{Total number of instances}} * 100\% \tag{3.1}$$

$$\text{Precision}(\%) = \frac{\text{True positive}}{\text{True positive} + \text{false Positive}} * 100\% \tag{3.2}$$

$$\text{Sensitivity}(\%) = \frac{\text{True positive}}{\text{True positive} + \text{false negative}} * 100\% \tag{3.3}$$

$$\text{Specificity}(\%) = \frac{\text{True negative}}{\text{True negative} + \text{false positive}} * 100\% \tag{3.4}$$

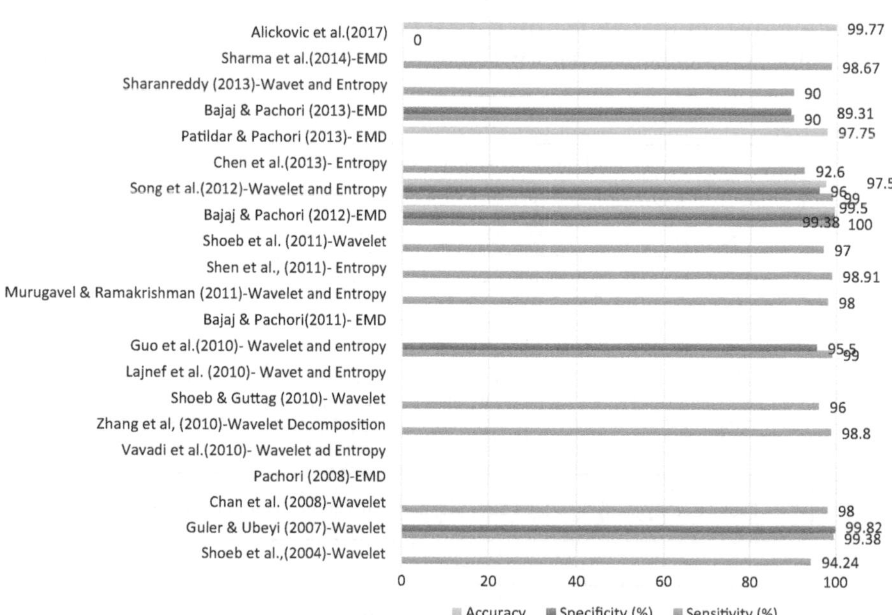

FIGURE 3.3
Previous studies results. (From Sylvia et al., 2019.)

In our study, a modified version of the original EMD algorithm [39] has been applied. This modified version [75] introduces novel criterias such as *mode amplitude* and *evaluation function*, which take predetermined thresholds into account to avoid over-iteration. Over-iteration often leads to over-decomposition of the original signal and can result in unwanted consequences, such as obliteration of physically meaningful amplitude fluctuations. However, the core principle of the EMD algorithm, i.e. that the EEG signals are broken down into a finite set of amplitude and frequency modulated components having zero local mean, is maintained. These components, termed as IMFs, are used to extract necessary features, which are ultimately fed to the classifier to distinguish between seizure and seizure-free data. An important point to note about the IMF components is that due to their non-harmonic nature, they are effectively able to capture the underlying morphologies of both the linear and non-linear components of the original signal in the time-frequency plane, as demonstrated by Rilling et al. [75] and depicted in Figure 3.4 [11]. The greater mathematical details of the algorithm can be found in Huang et al. [39].

We use the MATLAB™ code freely available at (http://perso.ens-lyon.fr/patrick.flandrin/emd.m), maintaining all the predefined threshold values ($\Theta_1 = 0.05$ and $\Theta_2 = 10\Theta_1$) for mode amplitude and evaluation function as described by Huang et al. [39] on our data. Other input parameters such as maximum number of iterations and interpolation scheme can be defined by the user in the code provided by (http://perso.ens-lyon.fr/patrick.flandrin/emd.m).

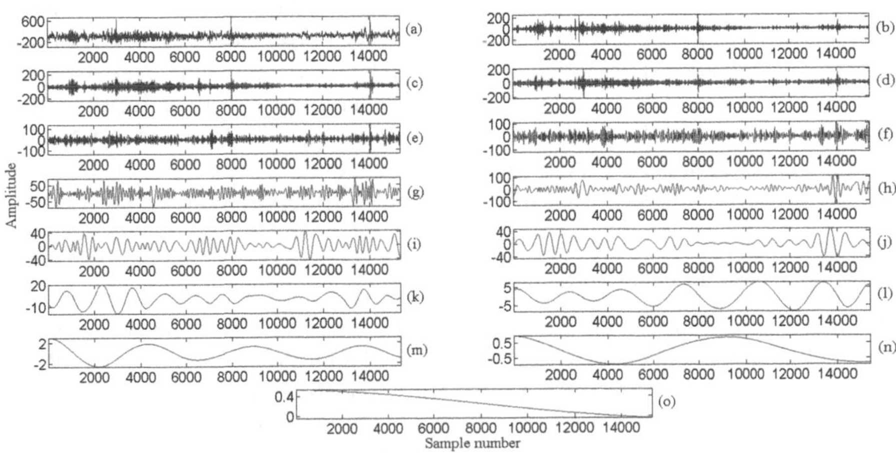

FIGURE 3.4
EMD-based analysis of one-minute seizure EEG signal in terms of IMF.

3.3.3 SVM Implementation

We classify the feature vectors with a two-class SVM classifier using radial basis function (RBF) kernel. Mathematically, the non-linear RBF function that classifies whether a feature vector X belongs to a seizure or non-seizure activity space is of the following form [41]:

$$\sum_{i=1}^{N} \alpha_i y_i e^{(-\gamma| |X-X_i| |^2)} + \beta > 0$$

As described by Huang et al. [41], a non-linear separable SVM consists of a hidden-layer perceptron with RBF nodes. Each RFB node has a centroid and impact factor, and its output is calculated by using a RBD of the distance between the input and the centroid factors.

The N coefficients of α_i, N support vectors of X_i, and the bias term (β) are determined by the SVM training algorithm, whereas the parameter γ is user-defined (y_i (0, 1)). It determines whether support feature vector X_i is a representative of seizure or seizure-free EEG. SVM classifier parameters and support vectors for seizure and seizure-free categories are computed offline during training time.

We use the concept of a k-fold cross-validation scheme with the parameter k set to 3. In this method, we randomly divide the seizure epochs in the dataset into three subsets. We then use two out of the three subsets as training dataset and the third one as test dataset to train the classifier. For each patient, we calculate the number of epochs correctly classified as seizure, number of false detections, and number of epochs of seizure data missed. This procedure was repeated three times such that each of the three subsets was used exactly once as test and twice as training dataset. For each seizure, the detection latency is calculated as the number of seizure epochs (in a span of the seizure duration) before the first seizure epoch that was correctly identified and classified during the electrographic onset of seizure. Averaging this count overall seizures gave the seizure latency for a particular patient.

A SVM consists of a single-layer feedforward network were an input feature vector is mapped into a higher dimensional space and, unlike in ANNs, the weights are adjusted only from the last hidden layer to the output layer. SVMs have been used in several studies [43, 67, 68] for epileptic seizure detection.

3.3.4 Performance Metrics

We use the standard performance metrics employed in seizure detection research defined in the study of Furbass et al. [30] for the representation of our results and performance comparison. The three standard parameters to measure the performance of a seizure detector are (a) Sensitivity, (b) Specificity, and (c) Latency. The performance of a seizure detector and the cost associated in obtaining that performance can be calculated from these parameters. Since our algorithm analyzes individual epochs discreetly, we categorize the epoch into a true positive detection (TP), a false positive detection (FP), a true negative detection (TN), or a false negative detection (FN).

3.4 Results and Discussion

We now discuss the results of our experiment in two different frameworks using the metrics discussed in Section 3.3.2. The results discussed in this section have been obtained using the mean frequency measure of the first and the last IMF components within a particular epoch as the input feature for classification.

3.4.1 Comparison with Studies Following Similar Performance Metrics

We now draw a comparison between the results of our study and that conducted by [66, 67]. The unique aspects of our study are (a) feature vector design and (b) the use of EMD algorithm instead of spectral methods for extracting features from EEG data. This study used a combination of spectral and spatial features to construct a feature vector space of M x N elements where *M* was the number of bandwidth filters used to extract spectral energies and *N* was the number of channels used to extract EEG data. Figure 3.5 represents our results graphically using the definition of standard performance metrics as defined in Equations 3.1–3.4.

Using Figure 3.5, we can evaluate and compare the total number of false positives obtained using the EMD method with that used in [66, 67]. We

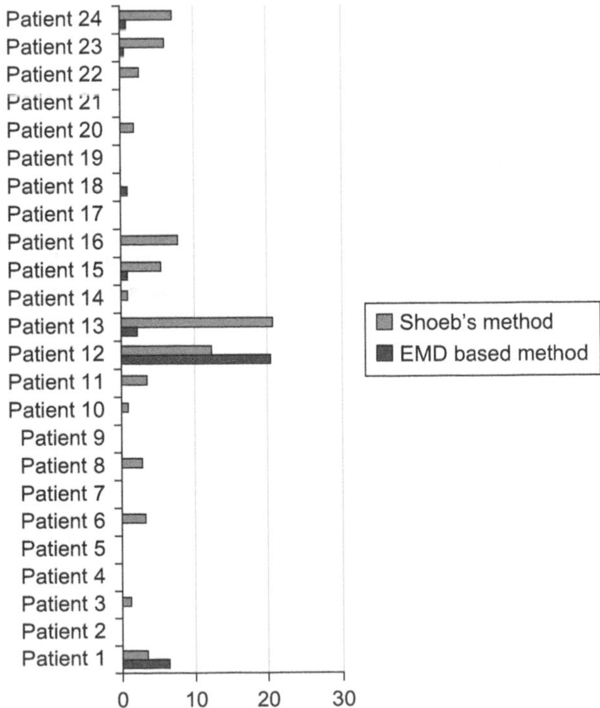

FIGURE 3.5
Average number of false detections in 24 hours.

found that our method gives a far lower number of false detections for 11 out of 18 patients (patient nos. 3, 6, 10, 11, 13, 15, 16, 20, 21, 23, 24), an equivalent number of false detections for 2 patients (patient nos. 4, 9), and inferior results for the remaining 5 patients (patient nos. 1, 2, 12, 18, and 19).

Previous studies [66, 67] trained the classifier on two or more seizures per patient and tested 916 hours of EEG data from 24 patients. Their proposed algorithm detected 96% of 173 test seizures with a median detection delay (latency) of three seconds, mean latency of 4.6 seconds and a median false detection rate of two false detections per 24-hour period. Our method trained three or more seizures per patient and tested 933 hours of continuous EEG data from 24 patients. The algorithm that we propose has median detection sensitivity of 100% with a median detection delay of 2.43 seconds and mean latency of 3.65 seconds. Hence, it is clear that, although our method's performance is slightly lower on the sensitivity front, it outperforms previous studies [66, 67] on specificity and latency measures. Another interesting way to compare the performance of the two methods is to look at the number of seizures correctly detected for each patient. Figure 3.6 represents this information.

In the case of the 17 patients in Figure 3.6, for which patient-specific data for sensitivity is available, we observe that our algorithm performs better for

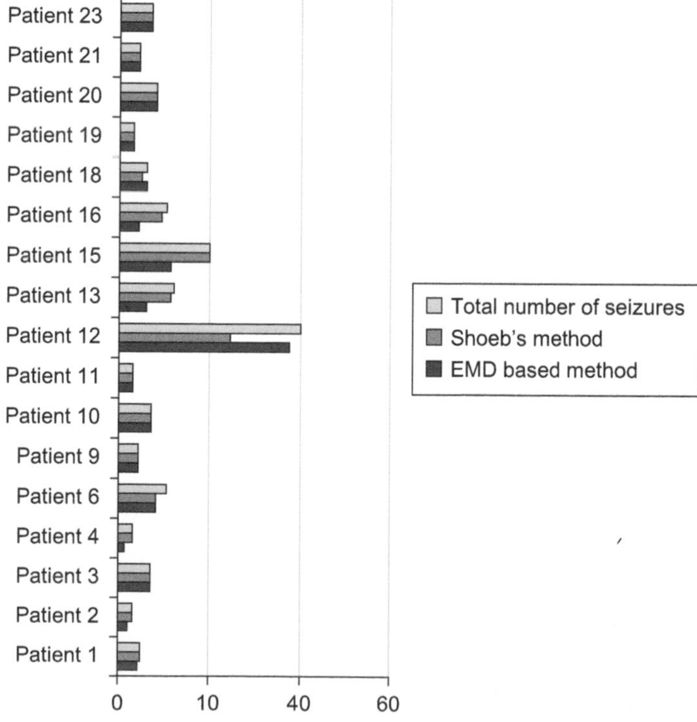

FIGURE 3.6
Number of seizures correctly detected per patient.

two patients (patient nos. 12 and 18), at par for 9 patients (patient nos. 3, 6, 9, 10, 11, 19, 20, 21, 23), and worse for 6 patients (1, 2, 4, 13, 15, 16) when compared to previous studies [66, 67].

3.4.2 Overall Performance

Table 3.1 contains results from individual patients for all the three-performance metrics discussed previously: specificity, sensitivity, and latency.

3.4.3 Using Epoch-Specific Values Instead of Event-Specific Values

One particular advantage of our proposed method is that we have calculated the epoch sensitivity in analyzing the data, whereas other studies [66, 67]

TABLE 3.1

Patient-Specific Results

ID	Hrs/Seizures/Epochs	Sensitivity (in Percentage)	False Detections (per 24 hrs)	Latency (in Seconds)
1	18/5/130	80	6.16	1.25
2	35.3/3/90	67	0.02	8.5
3	30/7/210	100	0.065	2.71
4	132/3/167	33.3	0	18
5	39/5/287	0	NaN*	NaN*
6	66.7/10/97	80	0.07	2.14
7	67/3/165	0	NaN*	NaN*
8	20/5/467	0	NaN*	NaN*
9	67.8/4/142	100	0.04	4.75
10	50/7/235	100	0.07	5.8
11	33.8/3/407	100	0.04	.67
12^	23.7/40/967	92.1	20.2	2.48
13^	33/12/278	50	2.1	3.67
14	36/8/97	0	NaN*	NaN*
15	39/20/743	55	0.9	4.8
16	19/8/39	42.5	0.2	1
17	20/3/91	0	NaN*	NaN*
18	34.6/6/164	100	0.9	1.9
19	28.9/3/121	100	0.1	2.3
20	27.6/8/159	100	0.007	1.25
21	32.8/4/106	100	0.005	2.5
22	31/3/108	0	NaN*	NaN*
23	26.5/7/159	100	0.4	1
24	21.3/16/157	100	0.63	2.38

* No seizures detected
** ID corresponds to patient number in the CHB-MIT database
^ Data for these patients were extracted using a variable number of electrodes/channels

have reported the event sensitivity. A study performed by Logesparan et al. [48] points out that the algorithms that consider event sensitivity report better results than epoch sensitivity, as decision thresholds are usually longer. For instance, a study that quotes 100% epoch sensitivity needs to detect all epochs in all seizures correctly, whereas a study that quotes 100% event sensitivity can have only one correctly detected epoch per seizure. Another inherent advantage in using epoch sensitivity instead of event sensitivity is that algorithms that report event sensitivity may not directly be compared if they use different detection durations. Hence, the way in which we represent our results is particularly handy while drawing out a comparison with similar studies.

3.5 Conclusion

The results prove that our method has substantial potential for use in a practical real-time seizure detector. In future, we plan to test EMD-based features on a one-class SVM to draw a parallel with the unary patient-specific detector described in Logesparan et al. [48]. We plan to experiment with the channel selection and feature ranking methods described and used by previous studies [9, 13, 21] to optimize the computational complexity of our method. Using these methods, we plan to optimize the time duration required for feature extraction and decrease the size of feature space used for classification to reduce computation costs. In our future experiments, we also propose tweaking the EMD algorithm to obtain a uniform number of IMF components within each epoch. Our purpose behind this is to standardize the feature space. In our future experiments, we plan to combine features from IMF components with features from electrocardiogram (ECG) data to verify whether this additional information improves detector performance.

References

1. H. Abdi and L. William, *Principal Component Analysis. Wiley Interdisciplinary Review: Computational Statistics*, Wiley Online Library, 2010.
2. B. Abibullaev, H. Seo, and W. Kang, "A wavelet based method for detecting and localizing epileptic neural spikes in EEG," *Science and Technology*, pp. 2–7, 2009.
3. A. Agrawal, L. Garg, and J. Dauwels, "Application of empirical mode decomposition algorithm for epileptic seizure detection from scalp EEG," in *The 35th Annual International Conference of the IEEE Engineering in Medicine and Biology Society (EMBC'13)*, 2013.

4. A. Ahmadi, M. Behroozi, V. Shalchyan, and M. R. Daliri, "Phase and amplitude coupling feature extraction and recognition of Ictal EEG using VMD," in *2017 IEEE 4th International Conference on Knowledge-Based Engineering and Innovation, KBEI* (Vol. 2018–January), 2018, pp. 0526–0532. Institute of Electrical and Electronics Engineers Inc. https://doi.org/10.1109/KBEI.2017.8325034.
5. R. Ahmed, A. Temko, W. Marnane, G. Boylan, and G. Lighbody, "Dynamic time warping based neonatal seizure detection system," in *Proceedings of the Annual International Conference of the IEEE Engineering in Medicine and Biology Society, EMBS*, 2012, pp. 4919–4922.
6. A. P. Aldenkamp, "Effect of seizures and epileptiform discharges on cognitive function," *Epilepsia*, vol. 38, no. s1, pp. S52–S55, 1997.
7. E. Alickovic, J. Kevric, and A. Subasi, "Performance evaluation of empirical mode decomposition, discrete wavelet transform, and wavelet packed decomposition for automated epileptic seizure detection and prediction," *Biomedical Signal Processing and Control*, vol. 39, pp. 94–102, 2018.
8. E. E. Audu, L. Garg, O. Falzon, and G. D. Giovanni, "Applications of machine learning in energy efficient, real time, monitoring, prediction, detection and management of seizure: localization of abnormal (seizure) EEG source," in *Mediterranean Neuroscience Society – 6th Conference*, St Julian's Malta, June 12–15, 2017.
9. M. Ayinala and K. K. Parhi, "Low complexity algorithm for seizure prediction using Adaboost," in *Proceedings of the Annual International Conference of the IEEE Engineering in Medicine and Biology Society, EMBS*, 2012, pp. 1061–1064.
10. S. Baillet, J. C. Mosher, and R. M. Leahy, "Mapping humain brain functions using intrinsic electromagnetic signals," Article invité dans *IEEE Signal Processing Magazine*, En Cours de Revision (Mai 2001), 2001.
11. V. Bajaj and R. B. Pachori, "Classification of seizure and nonseizure EEG signals using empirical mode decomposition," *IEEE Transactions on Information Technology in Biomedicine*, vol. 16, no. 6, pp. 1135–42, 2012.
12. V. Bajaj and R. B. Pachori, "Automatic classification of sleep stages based on the time-frequency image Of EEG signals," *Computer Methods and Programs in Biomedicine*, vol. 112, no. 3, pp. 320–328, 2013.
13. M. Bandarabadi, C. A. Teixeira, B. Direito, and A. Dourado, "Epileptic seizure prediction based on a bivariate spectral power methodology," in *Engineering in Medicine and Biology Society (EMBC), 2012 Annual International Conference of the IEEE*, 2012, pp. 5943–5946.
14. C. E. Begley, M. Famulari, J. F. Annegers, D. R. Lairson, T. F. Reynolds, S. Coan, and S. Dubinsky, "The cost of epilepsy in the United States: an estimate from population-based clinical and survey data," *Epilepsia*, vol. 41, no. 3, pp. 342–351, 2005.
15. D. Bhati, R. B. Pachori, and V. M. Gadre, "A novel approach for time-frequency localization of scaling functions and design of three-band biorthogonal linear phase wavelet filter banks," *Digital Signal Processing*, vol. 69, pp. 309–322, 2017.
16. A. Bhattacharyya and R. B. Pachori, "A multivariate approach for patient specific EEG seizure detection using empirical wavelet transform," *IEEE Transactions on Biomedical Engineering*, vol. 64, no. 9, pp. 2003–2015, 2017.
17. J. Bonello, L. Garg, G. Garg, and E. E. Audu, "Effective data acquisition for machine learning algorithm in EEG signal processing," in *Soft Computing: Theories and Applications*, 2018, pp. 233–244, Springer, Singapore.

18. M. R. Bouguelia, S. Nowaczyk, K. C. Santosh, and A. Verikas, "Agreeing to disagree: active learning with noisy labels without crowdsourcing," *International Journal of Machine Learning and Cybernetics*, vol. 9, no. 8, pp. 1307–1319, 2018.
19. S. Bugeja, L. Garg, and E. E. Audu, "A novel method of EEG data acquisition, feature extraction and feature space creation for early detection of epileptic seizures," in *38th Annual International Conference of the IEEE Engineering in Medicine and Biology Society (EMBC-2016)*, Orlando, FL, August 16–20, 2016.
20. S. Bugeja and L. Garg, "Application of machine learning techniques for the modelling of EEG data for diagnosis of epileptic seizures," in *The 3rd Workshop on Recognition and Action for Scene Understanding (REACTS 2015)*, Valletta, Malta, 2015.
21. N. Chang, T.-C. Chen, C.-Y. Chiang, and L.-C. Chen, "Channel selection for epilepsy seizure prediction method based on machine learning," in *Engineering in Medicine and Biology Society (EMBC), 2012 Annual International Conference of the IEEE*, 2012, pp. 5162–5165.
22. T. A. C. M. Claasen and W. F. G. Mecklenbrauker, "The Wigner distribution – a tool for time-frequency signal analysis. *Philips Journal of Research*, vol. 35, no. 3, pp. 217–250, 1980.
23. K. Cuppens, L. Lagae, B. Ceulemans, S. V. Huffel, and B. Vanrumste, "Detection of nocturnal frontal lobe seizures in paediatric patients by means of accelerometers: a first study," in *31st Annual International Conference of the IEEE EMBS*, Minneapolis, MN, 2009, pp. 6608–6611.
24. L. F. da Silva, *EEG: Origin and Measurement, EEG-fMRI. Physiological Basis, Technique and Application*, Springer, New York, NY, 2010, pp. 19–38.
25. L. J. Douglas, A high-resolution data-adaptive time-frequency representation. PhD dissertation, Rice University, 1987.
26. O. Falzon, K. P. Camilleri, and J. Muscat, "Complex-valued spatial filters for SSVEP-based BCI with phase coding," *IEEE Transactions on Bio-Medical Engineering*, vol. 59, no. 9, pp. 2486–2495, 2012.
27. H. Firpi, E. Goodman, and J. Echauz, "Epileptic seizure detection by means of genetically programmed artificial features," in *GECCO'05*, Washington, D.C., June 25–29, 2005.
28. R. J. Flanagan, R. A. Braithwaite, S. S. Brown, B. Widdop, and F. A. Wolff, *Basic Analytical Toxicology*, Word Health Organization, Geneva, Switzerland, 1995.
29. J. A. French, P. D. Williamson, V. M. Thadani, T. M. Darcey, R. H. Mattson, S. S. Spencer, and D. D. Spencer, "Characteristics of medial temporal lobe epilepsy: I. Results of history and physical examination," *Annals of Neurology*, vol. 34, no. 6, pp. 774–780, 1993.
30. F. Furbass, M. Hartmann, H. Perko, A. Skupch, P. Dollfuss, G. Gritsch, C. Baumgartner, and T. Kluge, "Combining time series and frequency domain analysis for a automatic seizure detection," in *Engineering in Medicine and Biology Society (EMBC), 2012 Annual International Conference of the IEEE*, 2012, pp. 1020–1023.
31. J.-M. Gerking, G. Pfurtscheller, and H. Flyvbjerg, "Designing optimal spatial filters for single trial EEG classification in a movement task," *Clinical Neurophysiology*, vol. 110, pp. 789–798, 1999.
32. A. Gevins, J. Le, N. Martin, P. Brickett, J. Desmond, and B. Reutter, "High resolution EEG: 124-channel recording, spatial deblurring and MRI integration methods," *Electroencephalography and Clinical Neurophysiology*, vol. 39, pp. 337–358, 1994.

33. A. L. Goldberger, L. Amaral, L. Glass, J. M. Hausdorff, P. Ivanov, R. G. Mark, J. Mietus, G. Moody, C. Peng, and H. E. Stanley, "PhysioBank, PhysioToolkit, and PhysioNet: components of a new research resource for complex physiologic signals," *Circulation*, vol. 101, no. 23, pp. e215–e220, 2000.
34. A. Guruvareddy and S. Narava, "Artifact removal from the EEG signals," *International Journal of Computer Applications*, vol. 77, no. 13, pp. 17–19, 2013.
35. T. Haddad, N. Ben-Hamida, L. Talbi, A. Lakhssassi, and S. Aouini, "Temporal epilepsy seizures monitoring and prediction using cross-correlation and chaos theory," *Healthcare Technology Letters*, vol. 1, no. 1, pp. 45–50, 2014.
36. J. D. Harel, "Complex principal component analysis: theory and example," *Journal of Climate and Applied Meteorology*, vol. 23, pp. 1660–1673, 1984.
37. H. Hassanpour, M. Mesbah, and B. Boashash, "Time-frequency feature extraction of newborn EEG seizure using SVD-based technique," *EURASIP Journal of Applied Signal Processing*, vol. 16, pp. 2544–2554, 2004.
38. L. Hlawatsch and B. Boudreaux, "Linear and quadratic time-frequency signal representations," *IEEE SP Magazine*, pp. 21–67, 1992.
39. N. E. Huang, S. Zheng, S. R. Long, M. C. Wu, H. H. Shih, Q. Zheng, N.-C. Yen, C. C. Tung, and H. H. Liu, "The empirical mode decomposition and the Hilbert spectrum for nonlinear and non-stationary time series analysis," *Proceedings of the Royal Society of London. Series A: Mathematical, Physical and Engineering Sciences*, vol. 454, pp. 903–995, 1998.
40. http://perso.ens-lyon.fr/patrick.flandrin/emd.m
41. K. Huang, D. Zheng, and I. King, "Arbitrary norm support vector machine," *Neural Computation*, vol. 21, pp. 560–582, 2009.
42. N.Kannathal, M. L. Choo, U. R. Acharya, and P. K. Sadaswan, "Entropies for detection of epilepsy in EEG," *Computer Methods and Programs in Biomedicine*, vol. 80, no. 3, pp. 187–194, 2005.
43. A. Karbouch, A. Shoeb, J. Guttag, and S. S. Cash, "An algorithm for seizure onset detection using intracranial EEG," *Epilepsy and Behaviour*, vol. 22, pp. S29–S35, 2011.
44. Z. J. Koles, M. S. Lazar, and S. Z. Zhou, "Spatial patterns underlying population difference in the background EEG," *Brain Topography*, vol. 2, no. 4, pp. 275–284, 1990.
45. Z. J. Koles, "The quantitative extraction and topographic mapping of the abnormal components in the clinical EEG," *Electroencephalography and Clinical Neurophysiology*, vol. 79, no. 6, pp. 440–477, 1991.
46. M. Koutroumanidis, S. Rowlinson, and S. Sanders, "Recurrent autonomic status epilepticus in Panayiotopoulos syndrome: video/EEG studies," *Epilepsy Behaviour*, vol. 7, no. 3, pp. 543–547, 2005.
47. J. Lachaux, D. Rudrauf, and P. Kahane, "Intracranial EEG and human brain mapping," *Journal of Physiology Paris*, vol. 97, pp. 613–628, 2003.
48. L. Logesparan, A. J. Casson, and E. Rodriguez-Villegas, "Improving seizure detection performance reporting: analysing the duration needed for a detection," in *Engineering in Medicine and Biology Society (EMBC), Annual International Conference of the IEEE*, 2012, pp. 1069–1072.
49. E. K. S. Louis and L. C. Frey, "Electroencephalography (EEG): an introductory text and atlas of normal and abnormal findings in adults, children, and infants." Retrieved from http://www.ncbi.nlm.nih.gov/pubmed/27748095, 2016.

50. H. Lu, H. L. Eng, K. N. Guan, K. N. Plananiotics, and A. N. Venetsanopoulou, "Regularized common spatial pattern with aggregation for EEG classification in small sampling setting," *IEEE Transactions on Biomedical Engineering*, vol. 57, no. 12, pp. 2936–2946, 2010.
51. H. Luders, J. Acharya, C. Baumgartner, S. Benbadis, A. Bleasel, R. Burgess, D. S. Dinner, A. Ebner, N. Foldvary, E. Geller, H. Hamer, H. Holthausen, P. Kotaga, H. Morris, H. J. Meencke, S. Noachtar, F. Rosenow, A. Sakamoto, B. J. Steinhoff, I. Tuxhorn, and E. Wyllie, "Semiological seizure classification," *Epilepsia*, vol. 39, no. 9, pp. 1006–1013, 1998.
52. H. Markandeya, S. Raghunathan, P. Irazoqui, and K. Roy, "A low-power 'near-threshold' epileptic seizures detection processor with multiple algorithm programmability," in *Proceedings of the 2012 ACM/IEEE International Symposium of Low Power Electronics and Design- ISLPED '12*, Redondo Beach, CA, July 30–August 1, 2012, p. 285.
53. R. J. Martis and U. R. Acharya, "Applications of higher order cumulant features for cardiac health diagnosis using ECG signals," *International Journal of Neural Systems*, vol. 23, no. 4, p. 13500142, 2013.
54. F. Masse, M. V. Bussel, A. Serteyn, J. Arends, and J. Penders, "Miniaturized wireless ECG monitor for real-time detection of epilepsy seizure," *ACM Transactions on Embedded Computing Systems*, vol. 12, no. 4, 2013.
55. F. Mormann, R. G. Andrzejak, C. E. Elger, and K. Lehnertz, "Seizure prediction: the long and winding road," *Brain*. Oxford University Press, 2007.
56. A. S. Muthanantha Murugavel and S. Ramakrishna, "An optimized extreme learning machine for epileptic seizure detection," *IAENG International Journal of Computer Science*, vol. 41, no. 4, pp. 212–221, 2014.
57. T. M. E. Nijsen, R. M. Aarts, J. B. Arends, and P. J. M. Cluitmans, "Model for arm movements during myoclonic seizures," in *Proceedings of the 29th Annual International Conference of the IEEE Engineering in Medicine and Biology*, 2007, pp. 1582–1585.
58. R. B. Pachori, "Discrimination between ictal and seizure-free EEG signals using empirical mode decomposition," *Research Letters in Signal Processing*, pp. 1–5, 2008.
59. R. B. Pachori and V. Bajaj, "Analysis of normal and epileptic seizure EEG signals using empirical mode decomposition," *Computer Methods and Programs in Biomedicine*, vol. 104, pp. 373–381, 2011.
60. E. Qazi, M. Hussain, H. Aboalsamh, W. Abdul, S. Bamatraf, and I. Ullah, "An intelligent system to classify epileptic and non-epileptic EEG signal," in *12th International Conference on Signal-Image Technology and Internet-Based Systems*, 2017, pp. 230–235.
61. Y. Qi, Y. Wang, X. Zheng, J. Zhang, J. Zhu, and J. Guo, "Efficient epileptic seizure detection by a combined IMF-VoE feature," in *Proceedings of the Annual International Conference of the IEEE Engineering in Medicine and Biology Society, EMBS*, 2012, pp. 5170–5173.
62. R. R. Sharma and R. B. Pachori, "Eigenvalue decomposition of Hankel matrix based time-frequency representation for complex signals," *Circuits Systems and Signal Processing*, vol. 37, no. 2, 2018.
63. J. W. Sander, "The problem of the drug resistant epilepsies, mechanism of the drug resistance in epilepsy," *Lessons From Oncology*, 2002.

64. R. R. Sharma and R. B. Pachori, "Improved eigenvalue decomposition of Hankel matrix based time-frequency representation for complex signals," *Circuits, Systems and Signal Processing*, vol. 37, no. 3, 2008.
65. M. Sharma and R. B. Pachori, "A novel approach to detect epileptic seizures using a combination of tunable-Q wavelet transform and fractal dimensions," *Journal of Mechanics in Medicine and Biology*, vol. 17, no. 7, 2017.
66. A. Shoeb, Application of machine learning to epileptic seizure onset detection and treatment. Ph.D. dissertation, MIT, September 2009.
67. A. Shoeb and J. Guttag, "Application of machine learning to epileptic seizures detection," in *Proceeding of the 27th International Conference on Machine Learning*, Haifa, Israel, 2010.
68. A. Shoeb, H. Edwards, J. Connolly, B. Bourgeois, S. Ted Treves, and J. Guttag, "Patient-specific seizure onset detection," *Epilepsy & Behavior*, vol. 5, no. 4, pp. 483–498, 2004.
69. N. Sriraam, S. Swathy, and S. Vijayalakshmi, "Development of a secure body area network for a wearable physiological monitoring system using a PSoC processor," *Journal of Medical Engineering and Technology*, vol. 36, no. 1, pp. 26–33, 2012.
70. W. Tatum, A. Husain, S. Benbadis, and P. W. Kaplan, *Handbook of EEG Interpretation. Medicine*, Demos Medical Publishing, LCC, 2008, p. 276.
71. M. Teplan, "Fundamentals of EEG measurement," *Measurement Science Review*, vol. 2, no. 2, pp. 1–11, 2002.
72. A. K. Tiwari, R. B. Pachori, V. Kanhangrad, and B. K. Panigrahi, "Automated diagnosis of epilepsy using key-point based local binary pattern of EEG signal," *IEEE Transactions on Biomedical Engineering*, vol. 21, no. 4, pp. 888–896, 2017.
73. T. Thayaparan, *Linear and Quadratic Time-Frequency Representation*. Defense Research Establishment, Ottawa, Canada, 2000.
74. H. Ramoser, J. Muller-Gerking, and G. Pfurtscheller, "Optimal spatial filtering of single trial EEG during imagined hand movement," *IEEE Transactions on Rehabilitation Engineering*, vol. 8, no. 4, pp. 441–446, 2000.
75. G. Rilling, P. Flandrin, and P. Gonçalvés, "On empirical mode decomposition and its algorithms," in *IEEE-EURASIP workshop on Nonlinear Signal and Image Processing, NSIP-03*, Grado, Italy, 2003.
76. S. Vajda, A. Karargyris, S. Jaeger, K. C. Santosh, S. Candemir, Z. Xue, S. Antani, and G. Thoma, "Feature selection for automatic turberculosis," *Journal of Medical Systems*, vol. 42, no. 8, 2018.
77. S. Vajda and K. C. Santosh, "A fast k-nearest neighbor classifier using unsupervised clustering," *Communications in Computer and Information Science*, vol. 709, pp. 185–193, 2017.

4

Fractal Analysis in Histology Classification of Non-Small Cell Lung Cancer

Ravindra Patil, Geetha M., Srinidhi Bhat, Dinesh M.S., Leonard Wee, and Andre Dekker

CONTENTS

4.1 Introduction

In both sexes combined, lung cancer is the most commonly diagnosed cancer (11.6% of the total cases) and the leading cause of cancer death. The total number of lung cancer cases in 2018 alone amounted to 2,093,876, the number of deaths with lung cancer being 1,761,007. Non-small cell lung cancer (NSCLC) accounts for 85% of all the lung cancers [1]. The cause of illness and the survival of NSCLC subjects vary across age, genetic profile, size of tumor, and histopathology of tumor. There are various studies that have established a correlation between the subtypes of NSCLC (squamous cell carcinoma, large cell carcinoma, adenocarcinoma, and "not otherwise specified") to the patient's survival. Also, it was studied that the prognosis for adenocarcinoma is poor compared to those for non-adenocarcinoma [2]. It was also concluded that surgical management should be different for each sub-category of NSCLC [3]. The current approach of subtype detection is performed using a biopsy procedure, where the tissue under observation is biopsied to determine the subtype, which is invasive in nature. The invasive approach is painful, costly, and not devoid of complications [4]. In recent times, several studies have been undertaken to identify the sub-categories of NSCLC non-invasively using radiomics, wherein large amount of quantitative features are mined and decision support models are built to achieve the desired objective [5]. Lately, radiomics has been applied to several medical problems such as tumors of lung, breast, and prostate, and also to images extracted from different medical imaging techniques (computed tomography (CT), magnetic resonance (MR), and positron emission tomography (PET)) [6–9, 10–13], showing promising results in each case.

Also, there has been lot of interest in the application of fractals in the oncology domain. Fractals are mathematical objects that have a non-integer dimension. These objects manifest a repeating pattern at different size scales; this property is quantified by a parameter named fractal dimension, which measures the self-similarity grade of the structure under analysis [14]. Such mathematical objects can be self-similar and resemble the repeated pattern within itself. These patterns have been studied in the oncology domain to differentiate between malignant and benign tumors in case of breast cancer. There has been comprehensive reviews on the use of fractal dimensions in various medical research areas, such as pathology, as shown by the literature [15, 16]. Recent trends show fractals to be a useful measure of the pathologies of the vascular architecture, tumor/parenchymal border, and cellular/nuclear morphology. A procedure that combines fractal and segmentation analyses has been proposed to investigate heterogeneity in cancer cells on MR images, similar to the approach described by Szigeti et al. for studying lung tumor heterogeneity on mice CT scans. More details on the fractals and their applications in oncology can be found in Baish and Rakesh [15].

Other major studies include different fractal measures, such as the power-law behavior of the Fourier spectrum of gray-scale images, which was used by Heymans et al. [17] to characterize the microvasculature in cutaneous melanoma. Fractal dimension quantified the degree of randomness to the vascular distribution, a characteristic that cannot be easily captured by the vascular density. Another study by Cusumano et al. aimed to use a fractal-based radiomic approach to predict the complete pathological response after chemotherapy in locally advanced rectal cancer (LARC). Fractal played an important role, giving information not only about the gross tumor volume (GTV) structure, but also about the inner populations in the GTV [18].

In this study, we investigate the role of fractals in the classification of NSCLC histology. We follow the box-counting approach, which aims to overlay boxes of various sizes over the area of interest and to optimize this quantity with the size of each box. Overall, our technical contributions are:

1. Building an algorithm which can compute the fractal dimension of a 2D region of interest in a given volume
2. Analyzing the extent to which fractal dimension aids in classification of NSCLC subtypes with the presence of other radiomic features

4.2 Methodology

The methodology adapted in this study consists of acquisition of NSCLC images with varied histology subtypes (i.e. squamous cell carcinoma, large cell carcinoma, adenocarcinoma, and "not otherwise specified"), through segmentation of the GTV, computation of fractals, extraction of radiomic features and validation of the model to identify the histology of the underlying tumor. The pictorial depiction of the workflow is shown in Figure 4.1

4.2.1 Image Analysis

The following experiment involved computerized tomography (CT) images of NSCLC. Images from 317 patients were used in this study, and the dataset was obtained from the collection of NSCLC radiomics in http://www.cancerimagingarchive.net/ [9]. The images were in digital imaging and communications in medicine (DICOM) format, and the CT image of each patient had the corresponding structural report file (RTSTRUCT), which consists of GTV delineation performed by a team of expert radiologists. The distribution of the demographic of data is depicted in Table 4.1.

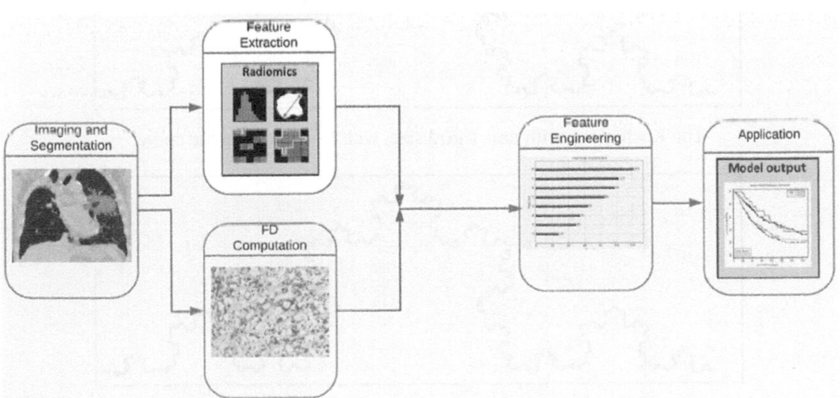

FIGURE 4.1
Adapted workflow and approach.

TABLE 4.1
Subject Demographics

Subject Characteristics	Adenocarcinoma	Large Cell Carcinoma	Squamous Cell Carcinoma	NOS
Number of subjects	40	108	110	59
Male	20	65	70	41
Female	20	43	40	18
Mean Age (years)	67.2	66.9	70.2	65.6

Accessing the DICOM tags, which contained the contour information of the GTV, using the information in the RTSTRUCT file, we created a mask depicting the GTV. The extracted mask was superimposed on the actual image to delineate the region of interest. Further, the minmax normalization of the images was performed to minimize the effects of spike pixels.

4.2.2 Computation of Fractal Dimension

The fractal dimension was computed based on the box counting approach and was further optimized to deduce the fractal dimension for each region of interest. In our approach, we extracted every surface contour of the GTV, and a box-counting algorithm was applied to compute the fractal dimension (FD) of each of the slices containing the tumor volume. The pictorial representation of the approach can be seen in Figure 4.2 with varied grid sizes [19].

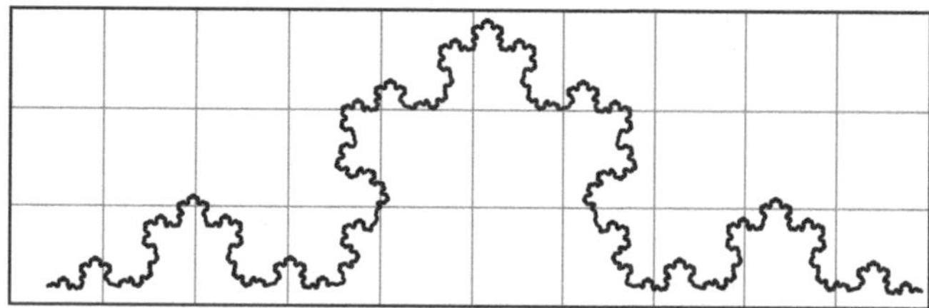

The Koch curve with unit 1 grid size, with 18 containing the curve.

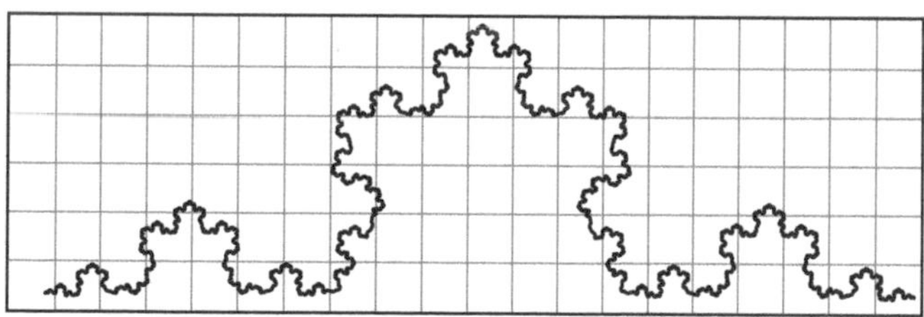

The Koch curve with unit 1/2 grid size, with 41 containing the curve.

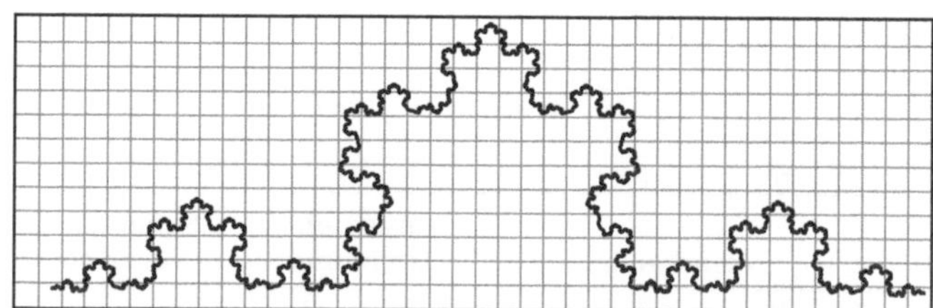

The Koch curve with unit 1/4 grid size, with 105 containing the curve.

FIGURE 4.2
Sample representation of box-counting approach with varied N [19].

In this study, the FD was computed on the impinged GTV on the CT slices. The equation for the computation is mentioned in (4.1).

$$FD = \frac{\log(N)}{\log\left(\frac{1}{r}\right)}, \tag{4.1}$$

where:

FD is the fractal dimension
N is the number of boxes needed to cover the region of interest
r is the size of each box

The above equation is recursively applied by varying the size of the box, thereby converting it into a curve-fitting solution. In principle, this corresponds to a line-fitting problem, where points corresponding to *N* and $1/r$ are fit and the slope of line provides the FD. This is one of the reasons we have applied a logarithm to Equation 4.1. This reduces a curve-fitting method to a line-fitting method, which is computationally simpler to solve. Also the logarithm, being a monotonic function in nature, doesn't alter the behavior of the original equation. The FD values for sample subsets are shown in Figure. 4.3, computed based on each of the slices.

4.2.3 Extraction of Radiomics Features

The quantitative image features were extracted from the GTV, wherein these imaging features where divided into four sub-categories: (1) first order statistics, (2) textural features, (3) shape- and size-based features, (4) wavelet features. The first order features provide the voxel intensity distribution within GTV. The textural features are computed using gray-level co-occurrence and gray-level run-length texture matrices, which aid in providing the relative position of various gray-level distributions. Shape and size features provide information on how spherical, elongated, or rounded the tumor is, and also about area, tumor compactness, and tumor volume. The wavelet features provide information by decoupling the region into high and low frequencies with GTV as input. In this approach, the original images are decomposed into eight levels of wavelet decomposition (X_{LLL}, X_{LLH}, X_{LHL}, X_{LHH}, X_{HLL}, X_{HLH}, X_{HHL}, and X_{HHH}), where L and H are low pass and high pass. For example, X_{LHL} is interpreted as the low-pass sub-band resulting from directional filtering in x with low pass, high pass along the y- direction, and low pass in the z-direction.

$$X_{LLH}(i,j,k) = \sum_{p=1}^{N_H}\sum_{q=1}^{N_L}\sum_{r=1}^{N_H} H(p)L(q)H(r)X(i+p,j+q,k+r), \tag{4.2}$$

where N_H is the length of filter H and N_L is the length of filter L.

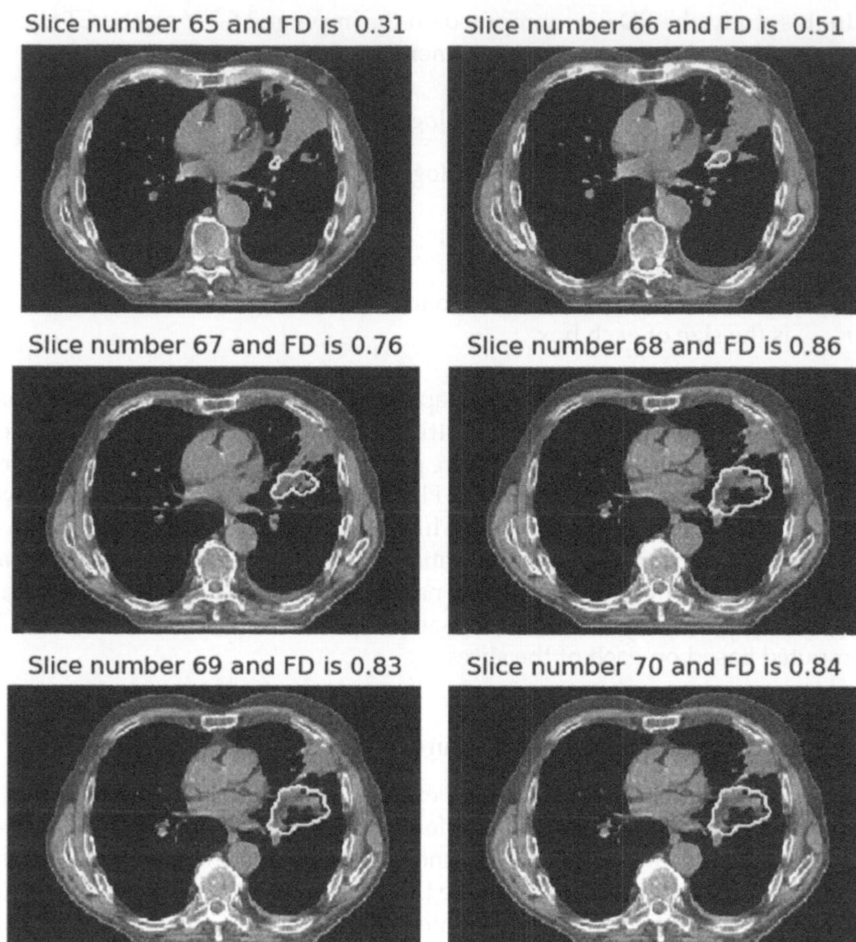
Slice number 65 and FD is 0.31
Slice number 66 and FD is 0.51
Slice number 67 and FD is 0.76
Slice number 68 and FD is 0.86
Slice number 69 and FD is 0.83
Slice number 70 and FD is 0.84

FIGURE 4.3
Subject Lung001 from the NSLC dataset with FD-computed value.

In total, 431 radiomics features were extracted, and these formed the feature vector for each of the subjects. Further, to this feature vector, max FD, average FD obtained from the fractal dimension computation was augmented, making it 433 feature vector for each subject.

4.2.4 Classification

Two data models where built in this experiment, one with all radiomic features, including extracted fractal features, and the other with only radiomic features, excluding the fractal features. A random forest classifier (RFC) was used to model the multiclass classification problem of predicting the histology of the tumor into one of the following sub-categories: squamous cell carcinoma, large cell carcinoma, adenocarcinoma, and "not otherwise specified." The RFC was implemented using the sklearn ensemble package in python and, to obtain the best set of parameters, we defined a grid of hyper-parameter ranges using the Scikit-Learn's RandomizedSearchCV package by performing 10-fold cross-validation with each combination of values. The hyper-parameters and the values of the RFC tuned in this experiment were:

- max_features: This parameter describes the maximum number of features random forest is allowed to try in an individual tree. The value in this experiment was chosen as "auto," which will take all the features in every tree. Though this option decreases the speed of the algorithm, it provides a high number of options to be considered at each node. Moreover, in this experiment we first chose the top 15 contributing features and then used them for the classification task, hence the sample space is reduced compared to the original 433.
- n_estimators: This parameter describes the number of trees to be built before taking the maximum voting of the predictions. Empirically, in this experiment, the optimal value for this parameter was 10.
- criterion: This parameter describes the criteria of split. In this experiment we have used the Gini index impurity measure.
- max_depth: This parameter represents the depth of each tree in the forest. The deeper the tree, the more splits it has, and the more information it can captureregarding the data. In this case, the optimal depth value was found to be 15.
- min_samples_leaf: This describes the minimum samples required to be at the leaf node. In this experiment, the optimal value of the parameter was found to be 3.

4.2.5 Results

A total of 317 subjects were considered for the histology classification, out of which 40% of subjects were female, the average age being 68 years. Pearson's correlation analysis was performed to understand the relationship of first order radiomics and FD features with the histology class. It can be observed that max FD has maximum correlation with respect to histology class compared to other first order features. Further, it can also be seen that tumor volume as an independent feature ranks much lower than the FD in Figure 4.4. This is in line with the understanding that the morphology of the tumor provides a more vital distinction than does its volume in histology differentiation.

The classification accuracy considering the radiomics features accounted for 74%; however, with the inclusion of FD features, it increased to 86% ($P < 0.001$). Also, there is an improvement of 8% and 17% in terms for sensitivity and specificity respectively by considering the FD features (Figure 4.5).

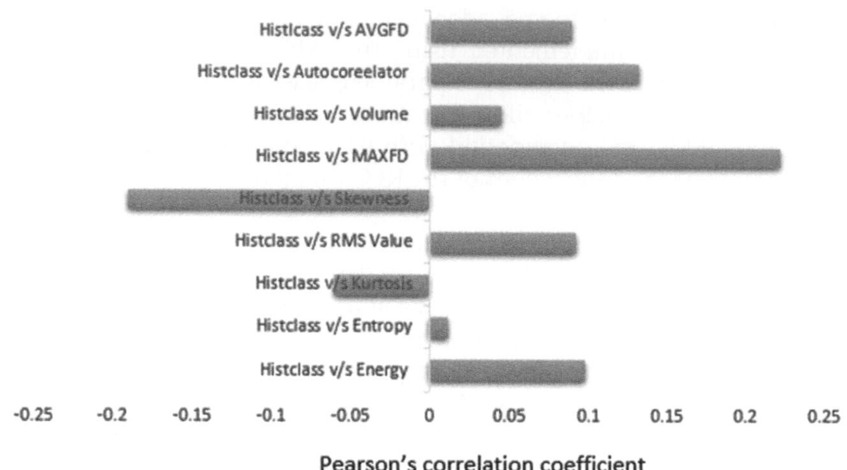

FIGURE 4.4
Correlation of histology with various first order radiomics features and FD features.

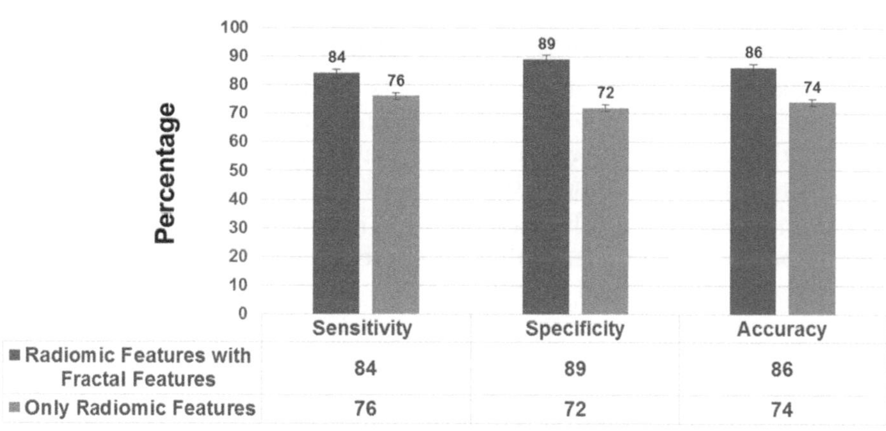

	Sensitivity	Specificity	Accuracy
Radiomic Features with Fractal Features	84	89	86
Only Radiomic Features	76	72	74

FIGURE 4.5
Classification metric comparison between with-FD and without-FD features inclusion.

Further, the features were ranked based on the random forest classifier on the priority of importance for classification of histology, as shown in Figure 4.6. The top features contributing toward histological classification using radiomic features include HHH_Sum Entropy, HHH_LRGE, HHH_RLN, inverse difference moment normalized (IDMN), HHH_GLN, HHH_SRHGE, HHH_IDMN, informational measure of correlation 2 (IMC2), maximum FD, average FD, HHH_Sum Variance. Also, it's interesting to note

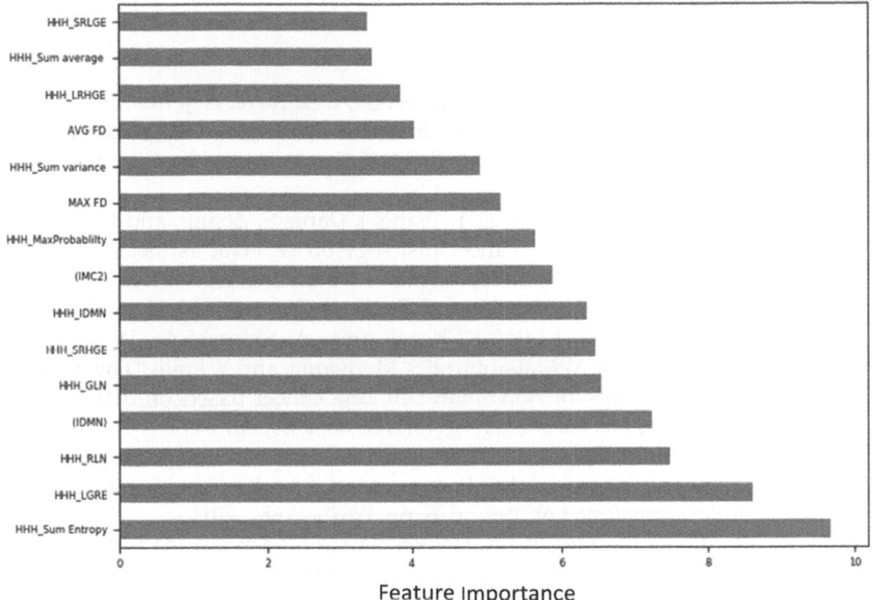

FIGURE 4.6
Feature ranking based on importance.

that the FD-derived, max FD, and avg FD features were ranked among the top 15 features that aid in histology classification of NSCLC. In essence, wavelet-based features and FD parameters dominate as top contributing features for histology classification.

4.3 Conclusion

In this study, we established that FD features play an important role in histology classification of NSCLC. Applying fractal analysis on a 2D contour region can provide valuable information and reflect the idea of overall tumor aggressiveness. However, our study has a limitation: the fractal computation was applied on 2D contour GTV regions. This study could be extended to a 3D fractal analysis algorithm using the mesh approach, so that the region of GTV would be better delineated for computation FD. Also, for a tumor contour that is very small, there might not be enough points in the number of boxes vs. the size of each box plane ($\log(N)$ vs. $\log(1/r)$) to describe the tumor contour. In this case, the best fit line will be only an approximation, and , this will be reflected in the FD value as well.

References

1. F. Bray, J Ferlay, I Soerjomataram, R.L. Siegel, L.A. Torre, A. Jemal, "Global cancer statistics 2018: GLOBOCAN estimates of incidence and mortality worldwide for 36 cancer in 185 countries," *CA: A Cancer Journal for Clinicians*, vol. 0, pp. 1–31, 2018.
2. L. H. Ma, G. Li, H. W. Zhang, Z. Y. Wang, J. Dang, S. Zhang, and L. Yao, "The effect of nonsmall cell lung cancer histology on survival as measured by the graded prognostic assessment in patients with brain metastases treated by hypofractionated stereotactic radiotherapy," *Radiation Oncology*, vol. 11, p. 92, 2016.
3. M. Yano, J. Yoshida, T. Koike, K. Kameyama, A. Shimamoto, W. Nishio, K. Yoshimoto, T. Utsumi, Y. Matsumura, S. Moriyama, and Y. Fujii, "The outcomes of a limited resection for non small cell lung cancer based on differences in pathology," *World Journal of Surgery*, 40(11):2688–2697, 2016.
4. C. C. Wu, M. M. Maher, and J. A. Shepard, "Complications of CT-guided percutaneous needle biopsy of the chest: prevention and management," *American Journal of Roentgenology*, vol. 196, no. 6, pp. W678–W682, 2011.
5. R. Patil, G. Mahadevaiah, and A. Dekker. "An approach toward automatic classification of tumor histopathology of non–small cell," *Tomography*, 2(4):374–377, 2016.
6. R. Gillies, P. Kinahan, H. Hricak, "Radiomics: images are more than pictures, they are data," *Radiology*, vol. 278, p. 2, 2016.
7. P. Lambin, E. Rios-Velazquez, R. Leijenaar *et al.* "Radiomics: extracting more information from medical images using advanced feature analysis," *European Journal of Cancer*, vol. 48, pp. 441–446, 2012.
8. V. Kumar, Y. Gu, S. Basu, *et al.* "Radiomics: the process and the challenges," *Magnetic Resonance Imaging*, vol. 30, no. 9, pp. 1234–1248, 2012.
9. H. J. Aerts, E. R. Velazquez, R. T. Leijenaar, *et al.* "Decoding tumour phenotype by noninvasive imaging using a quantitative radiomics approach," *Nature Communications*, vol. 5, p. 40, 2014.
10. K. C. Santosh and Sameer K. Antani, "Automated chest X-ray screening: can lung region symmetry help detect pulmonary abnormalities?" *IEEE Transactions on Medical Imaging*, vol. 37, no. 5, pp. 1168–1177, 2018.
11. S. Vajda, A. Karargyris, S. Jäger, K. C. Santosh, S. Candemir, Z. Xue, S. K. Antani, and G. R. Thoma, "Feature selection for automatic tuberculosis screening in frontal chest radiographs," *Journal of Medical Systems*, vol. 42, no. 8, pp. 146:1–146:11, 2018.
12. F. T. Zohora and K. C. Santosh, "Foreign circular element detection in chest X-rays for effective automated pulmonary abnormality screening," *International Journal of Computer Vision and Image Processing*, vol. 7, no. 2, pp. 36–49, 2017.
13. K. C. Santosh, Szilárd Vajda, Sameer K. Antani, and George R. Thoma, "Edge map analysis in chest X-rays for automatic pulmonary abnormality screening," *International Journal of Computer Assisted Radiology and Surgery*, vol. 11, no. 9, pp. 1637–1646, 2016.
14. B. B. Mandelbrot, *The Fractal Geometry of Nature*, W. H. Freeman, New York, 1982.
15. J. W. Baish and R. K. Jain, "Fractals and cancer," *Perspectives in Cancer Research*, vol. 60, no. 14, 2000.

16. I. Sokolov, "Fractals: a possible new path to diagnose and cure cancer?" *Future Oncology*, vol. 11, no. 22, pp. 3049–30513, 2015.
17. O. Heymans, S. Blacher, F. Brouers, and G. E. Pierard, "Fractal quantification of microvasculature heterogeneity in cutaneous melanoma," *Dermatology*, vol. 198, pp. 212–217, 1999.
18. D. Cusumano *et al. Development and Validation of New Radiomic Features Based on Fractal Analysis.*
19. Fractal Explorer. "Calculating fractal dimension." Availale at: https://www.wahl.org/fe/HTML_version/link/FE4W/c4.htm, last accessed and verified on November 4, 2018.

5

Multi-Feature-Based Classification of Osteoarthritis in Knee Joint X-Ray Images

Ravindra S. Hegadi, Dattatray N. Navale, Trupti D. Pawar, and Darshan D. Ruikar

CONTENTS

5.1 Introduction

Osteoarthritis (OA) is a type of arthritis that may occur due to breakdown and subsequent loss of a portion of cartilage in the joints. Cartilage is formed of protein substance and works as a cushion in between the bones of a joint. Due to the loss of cartilage, two bones can rub against each other causing swelling, pain, and difficulty in the motion of joints. Gradually, the shape of the bone will also be lost.

Further, it may cause the growth of bone spur at the edges of the bone joint. In severe cases, the bone or cartilage may break into pieces and float inside the joint region resulting in severe pain and damage. The effect of pain is higher with an increase in the activities of the affected part, and lower when the part is rested. OA is one of the more than 100 different types of conditions of arthritis. The most common locations of occurrence of OA are the joints of hands, feet, spine, hips, and knees. Arthritis is known as "primary" if the reasons for the occurrence of arthritis are not known; if they are, then it is called "secondary arthritis" [1]. Elderly persons are the most affected by OA. As per the statistics available from the World Health Organization (WHO), among people above 60 years, 9.6% of men and 18% of women are affected by OA diseases across the world. Among rheumatologic problems, OA is the second most occurring disease, with 22% to 39% among all the rheumatologic problems in India. About 45% of Indian women above the age group of 65 years are suffering from OA.

5.2 Causes of OA

There are many factors for the occurrence of OA. With the increase in age, OA also progresses until the age of 80 years, with women being more affected compared to men. Traumas such as meniscal tears, injury to a collateral ligament, and fractures in joints may also lead to OA. Excessive usage of the knee part due to heavy physical tasks in routine jobs and repeated use of a joint in regular work, for instance in the case of physical instructors, could lead to a higher risk of developing OA. Athletes too have a high probability of developing OA. Obese people around 35 years of age have a higher risk of OA because, due to overweight, the possibility of a cartilage tear is very high. Gender-wise, women have a high risk of developing OA compared to men, specifically in European countries [2].

5.3 Levels of Knee OA

The proposed work classifies normal and OA-affected knee images, with a detailed study on the grading of knee OA. There are five stages of knee OA based on the severity of disease progression, as proposed by the Kellgren-Lawrence Grading Scale [3]. In stage 0, also known as the normal or healthy knee, the knee shows no symptoms of OA. The patients with stage 1 OA, which is a mild stage, will have mild wear and tear of meniscus, and evidence of bone spur growths at the ends of knee joints may be found. This also causes minor narrowing of joint space along with possibilities of osteophytic lipping. The X-ray of stage 2 knee OA shows definite evidence of osteophytes and visible narrowing of joint space. In this stage, more evidence of bone spur growth is visible, and patients will experience symptoms of joint pain. The knee joint becomes stiff and uncomfortable during a more extended period of sitting or sleeping. In this stage, there will be a proteolytic breakdown of the cartilage matrix due to the increased production of enzymes, but cartilage and soft tissues in the knee part will still be healthy. In stage 3, a moderate stage, there will be clear evidence of moderate multiple osteophytes, and limited visibility of narrowing of joint space will be found. This also causes some sclerosis and possible deformity of bone contour. The gap between bones is due to the erosion of the cartilage surface. With the progress of the disease, proteoglycan and collagen fragments are released into the synovial fluid. This situation leads to the development of spurs in bone joints; moreover, bone joints become rougher. Patients with stage 3 OA suffer from joint inflammation leading to frequent pain during walking. Popping and snapping sounds may also be heard while walking. Stage 4 OA is defined as severe. In this stage, the gap between the knee bones is considerably reduced, large osteophytes are found, severe sclerosis is developed, and definite deformity of bone contour will be visible. In this stage, most of the cartilage has been lost due to wear, and the knee joint has become stiffer. Due to a reduced level of synovial cavity fluid, friction between bones happens, causing severe pain while walking or moving the joint. In this stage, synovial metalloproteinase, cytokines, and tumor necrosis factor (TNF) are excessively produced, destroying soft tissues around the knee. The X-ray images of different stages of OA are shown in Figure 5.1.

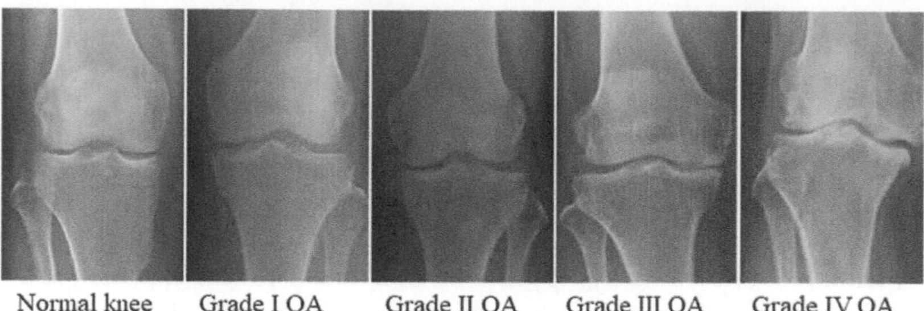

FIGURE 5.1
Grades in OA. (Image courtesy of Ju Hee Ryu et al. [4].)

5.4 Proposed Work

Radiography, through tools such as X-ray, magnetic resonance imaging (MRI), and computed tomography (CT) scan images, helps physicians decide on the possibility of OA along with the other symptoms. The level of severity of OA is presented in the form of grades. One such grading, presented by Kellgren and Lawrence (K&L), is the most accepted and used classification in the identification of OA by radiologists across the world [5]. According to K&L, OA has five grades that go from 0 to 4 as described in the previous sub-section. Several researchers used the K&L criteria for classification and grading of OA in their research work. The same criteria were adopted by the WHO for their epidemiological studies of OA. In their first publication regarding OA in 1963, the description of grading was not provided. Many studies were published on later days using original K&L criteria, but there were differences in the descriptions and their impact on classification and distribution of severity of knee OA in these studies. Further, these studies could not establish a perfect association between clinical knee complaints and grading criteria.

As per the description of grading, there is a direct association between the pain in the knee part of the patient suffering from OA and the severity of knee OA. A patient with a higher-grade OA is expected to experience more knee pain. However, in a practical situation, people suffering from moderate (grade 3) to severe (grade 4) knee OA can be found to have no knee pain. Why they are not experiencing pain despite having such a high grade of knee OA is not known.

The visual analysis of radiograph images has mostly followed the methodology in assessing the OA grade. Among different available radiography methods, MRI is the most useful since it provides a way to assess the cartilage part of the knee, which plays a vital role in grading knee OA. It also provides information on structural damage to the cartilage. However, MRI is an expensive procedure and will not be commonly available, specifically in rural and semi-urban locations. A cost-effective and readily available alternative radiographic technique to MRI is X-ray. In this chapter, different methods are proposed for the classification of a knee joint as normal or abnormal (i.e. affected with OA). The images are preprocessed using different preprocessing techniques and curvature, and texture features are extracted. Multiple classifiers are used to classify the images and results are compared.

5.5 Literature Survey

Due to technological advancements, the orthopedic healthcare field is taking the initiative to adopt computerized solutions. In the past, many researchers worked to develop virtual reality-based simulators for skill training, automated preoperative surgical planners, outcome prediction systems, and interpretative assistance systems [6]. In addition to this, several successful attempts were made to develop a computer-assisted diagnosis (CAD) system that grades levels in knee OA using X-ray, CT scan, and MRI images. Different preprocessing techniques, segmentation methods, and classifiers were used by these researchers to detect and segment the synovial cavity region, to identify whether the image is normal or abnormal, and to grade these images.

R. S. Hegadi et al. [7] identified knee OA using a block-based texture analysis approach and a support vector machine (SVM) classification technique. In their work, they divided images into nine equally sized blocks and extracted different texture features, namely, skewness, kurtosis, standard deviation, and energy from each of these nine blocks, leading to 36 feature vectors for each image. An SVM classifier was used to classify images as normal and affected. The accuracy of their algorithm is 80% for normal images and 86.7% for the images affected by OA.

J. Duryea et al. [8] developed a semi-automated method for measuring the loss of knee cartilage in patients suffering from OA. Variations in the cartilage volume in five regions were recorded over a period of time. Measures such as average change in volume, standard deviation, and standardized response time were used for their study. Further, they used a 3D coordinate system based on cylindrical coordinates for measuring the change in cartilage volume. By inspecting 24 patient-specific studies, fixed measurement locations were determined to segment synovial cavity region. In this case, if the subjects and time points are not the same, then the comparison with the other studies is critical.

Classification of OA images using a self-organizing map was proposed by L. Anifahet et al. [9]. This work adopted contrast limited adaptive histogram equalization (CLAHE) and template matching to decide whether an image belongs to the left or right knee. To identify the synovial cavity region, they segmented the knee image using the Gabor kernel, template matching, row sum graph, and gray-level center of the mass methods. Gray-level co-occurrence matrix (GLCM) features are used for further classification of data in five grades of OA. Result for images of grade 0 and grade 4 were found to have a higher rate of accuracy than images from grade 1 to grade 3. Accuracy, specificity, and sensitivity were the features used for the evaluation of the outcome.

Automatically quantifying radiographic knee OA severity using deep convolution neural networks (CNN) was proposed by Joseph Antony et al. [10]. Here, assessment of knee OA severity has been approached as an image classification problem with K&L grades being ground truth for classification. The authors used bilateral posteroanterior (PA) fixed flexion knee X-ray images taken from the baseline (image release version O.E.1) containing radiographs of the Osteoarthritis Initiative (OAI) dataset obtained from 4,476 participants for their experimentation. A linear SVM classifier with five-fold cross validation produced 95.2% accuracy.

Fully automated, level-set-based segmentation for knee MRIs using an adaptive force function and template data from the OAI was proposed by Chunsoo Ahn et al. [9, 11]. They adopted two methods, namely, level-set-based segmentation and new template data. The experimental results were evaluated using dice similarity coefficients (DSCs), which showed 87.1%, 84.8%, and 81.7% of accuracy for femoral, patellar, and tibial cartilage respectively. For experimentation, data from 10 subjects were used. Most of the research attempts found in the literature considered only two types of knee cartilage tissues (femoral and tibial) for identification of OA; however, for this experimentation, three types of knee cartilage tissues (femoral, patellar, and tibial) were used.

The use of textural and statistical features such as entropy, standard deviation, coarseness, and contrast for analyzing the severity of radiographic OA of the knee joint was proposed by Pooja P. Kawathekar et al. [12]. These features demonstrated variations depending upon the severity of knee OA. For classification purposes, they use the grade methods that are assigned to knee-joint images according to their synovial cavity. Their proposed algorithm could detect and classify OA of the knee joint with sufficient accuracy. This methodology demonstrated a high accuracy of 92% for grade 3 images.

A fully automatic knee OA CAD system for the quantification of major OA parameters on plain knee radiographs was developed by H. Oka et al. [13]. For this research study, the authors developed a fully automatic program known as knee OA computer-aided diagnosis (KOACAD). KOACAD is used to quantify OA, which was further validated for reproducibility and reliability of their system. For further research, in 1979, some anteroposterior radiographs of knees of a large-scale cohort population were utilized. The data were analyzed by KOACAD as well as by conventional categorical grading systems.

A novel method to segment the synovial cavity region from knee X-ray images is presented by R. S. Hegadi et al. [14]. In the proposed methodology, the OA images were preprocessed using the Wiener filter and thresholding technique. After preprocessing, the synovial cavity region was located by identifying the distribution of intensity. After finding the synovial cavity region, they segmented this region from knee X-ray images. Finally, the authors compared their results with the manually segmented images and found that their results were in agreement with the manual segmentation results.

A scheme to segment healthy knees from clinical MRI images was proposed by Jurgen Frippet et al. [15]. Their work consisted of three stages:

segmentation of bone region, extraction of the interface between bone and cartilage, and finally cartilage extraction. For segmentation of bone, they used 3D active shape models. For the extraction of the bone and cartilage interface, prior knowledge based on points belonging to the region was used.

Segmentation of cartilage from knee MRI images using the watershed algorithm was proposed by Patel et al. [16]. Watershed segmentation involves great simplicity of application. For early detection of OA and segmentation, they use MRI images of the knee cartilages. After detection of OA, they need to segment the synovial cavity region. So, for segmentation purposes, MRI DICOM (digital image and communication in medicine) images were taken. Then they overlaid the detected region into the original knee MRI DICOM image.

A deep learning-based spproach for automatic knee OA diagnosis from plain radiographs was proposed by Aleksei Tiulpin et al. [17]. They proposed a new technique called transparent computer-aided diagnosis method, based on the deep Siamese CNN, to automatically score knee OA severity according to the K&L grading scale. The dataset from the multicentre OA study was used and validated on 3,000 randomly selected subjects (5,960 knees) from the OAI dataset to train that dataset properly. Their proposed method yielded a quadratic kappa coefficient of 0.83 and average multiclass accuracy of 66.71% compared to the annotations given by a committee of clinical experts. Here, they also reported on a radiological OA diagnosis area under the ROC curve of 0.93.

Detection of knee OA using X-ray images was proposed by Mahima Shankar Pandey et al. [18]. X-ray images will not provide more details on the exact location of injured bone when compared to MRI and CT images, which is a challenge. For this work dataset on the knee, X-ray images were obtained from the Department of Medical Informatics, Aachen University of Technology, Germany. The dataset named IRMA has 10,000 images organized in 57 categories collected in 2005. These images were preprocessed with the histogram equalization and contrast stretching methods to enhance them. Further challenges were the detection of an edge on preprocessed images and cropping only the synovial cavity region. To achieve better results, binary operations were applied. Since the boundary of each bone was known, it was used to determine the thickness in order to detect OA.

Texture analysis using the complex wavelet decomposition for knee OA detection was proposed by Rabia Riad et al. [19]. They use the classification of subjects with different stages of knee OA using bone texture analysis. The proposed research work is mainly subdivided into five parts. An image preprocessing step is introduced to keep the crucial information of bone texture. After the preprocessing of images, decimated dual-tree complex wavelet transform is employed on the filtered ROIs. Complex wavelets coefficients are adopted and modeled using the von Mises and wrapped Cauchy distributions employing some parameters. The parameters for each model were estimated using the maximum likelihood estimator algorithm, and, further, these parameters were used in the classification of the bones. Finally, the performance of the proposed method was evaluated using the knee radiographs.

5.6 The Proposed Methodology

In this work, two methods are employed for the classification of knee X-ray images as normal or affected by OA. Namely, curvature-based and texture-based classification. The results obtained are compared.

5.6.1 Noise Removal and Image Enhancement

At first, the patient-specific X-ray images having synovial cavity regions are accepted. They are in RGB color format. Before proceeding to further processing, these images are converted to a gray-scale format. Figure 5.2(a) shows the image in RGB format whereas Figure 5.2(b) shows the converted image in gray-scale format. X-ray images are susceptible to noise. The noise may be added during image capturing, or because of the poor quality of the camera; there might also be some environmental effects, such as high temperature or humidity. To remove noise and enhance the desired portion (synovial cavity region), it is necessary to apply effective noise removal and image enhancement techniques [24, 28]. There are many types of noise removal techniques, such as order filter, median filter, Gaussian filter, and order statistics filters; however, in the proposed work we have adapted the Wiener filter to remove noise. It is a 2D adaptive noise-removal filter that adapts itself for local variance in the image [20]. There are five primary reasons behind the selection of the Wiener filter. These are as follows: (a) For higher local noise, a variance filter performs a lesser amount of smoothing and, for lower local noise, a variance filter performs a greater amount of smoothing. (b) The performance of this filter is better as compared to the linear filters. (c) The filter is quite efficient in the time required for computation as compared to linear filtering. (d) It performs best for white additive noise such as Gaussian noise. (e) The mean square error of this filter is optimal, which means the mean square error is minimized during filtering and noise-reduction process. The Wiener filter is optimal in terms of the mean square error. After filtering, the contrast of the filtered image is

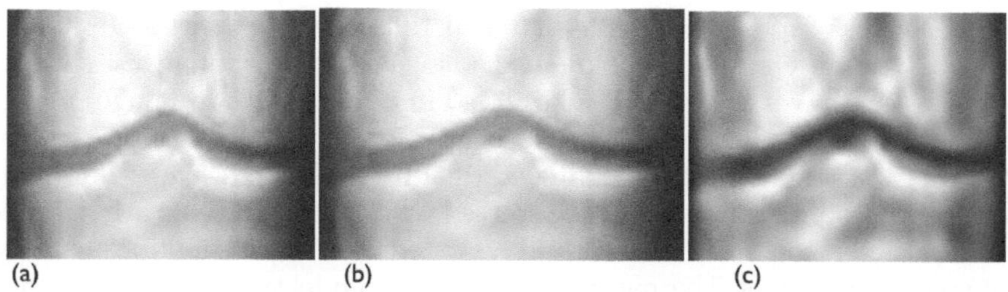

FIGURE 5.2
Stages of image preprocessing: (a) Original image, (b) Converted to gray-scale, and (c) Image after noise removal.

enhanced using CLAHE [21]. Through this method, the values of image pixel intensities are enhanced by operating into smaller regions rather than the entire image. The neighbouring enhanced regions are joined using a bilinear interpolation technique to avoid generation of boundaries for the processed blocks. The block size chosen for applying CLAHE is 8 × 8. The result of these enhancement processes is shown in Figure 5.2(c).

5.6.2 Curvature-Based Feature Extraction Method

In this method, an initially preprocessed image is accepted to extract the synovial cavity region. To achieve this, we used an active contour model. After segmentation, the boundaries of the synovial cavity region are obtained using morphological erosion technique with 3 × 3 structuring element. The features used for classification are the curvature values obtained over these edges. Pixel gradients are used for obtaining the curvature gradient matrix along the x and y axes. Normal images generally have smooth curvature values as compared to images with OA. The mean and standard deviation values are obtained and the k nearest-neighbor classifier is used for classification. The following sections describe in detail the proposed methodology.

5.6.3 Segmentation of Image

After noise removal, the active contour-based segmentation method [22, 30, 31] is applied to separate the synovial region from the background. For the implementation of the active contours, a region-based energy model is used, which is an approximation for a multistage segmentation model. This multistage model uses level sets for the formulation of curve evolution, as shown in Figure 5.3. The level set has an implicit contour in which the evolving curve is treated at the zero-level line of the level set function. The basic purpose of the active contour technique is to segment the input image into two

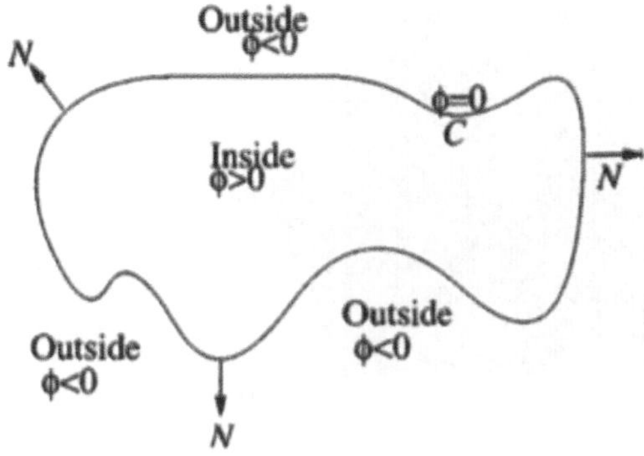

FIGURE 5.3
Formulation of curve evolution using the level set method.

regions known as object region and background, by embedding the boundary of an object using the zero-level curve of the 3D level set function. The original image and the results of segmentation using active contours are shown in Figure 5.4 (a) and (b) respectively.

5.6.4 Boundary Extraction

The segmented image is in binary form. Simple morphological boundary extraction using a 3 × 3 structuring element is applied to extract the inner boundary of the segmented region, which separates the synovial cavity region from the bone parts. The result of the extracted boundaries is shown in Figure 5.5 (a) and the boundaries in the original image are shown in Figure 5.5 (b).

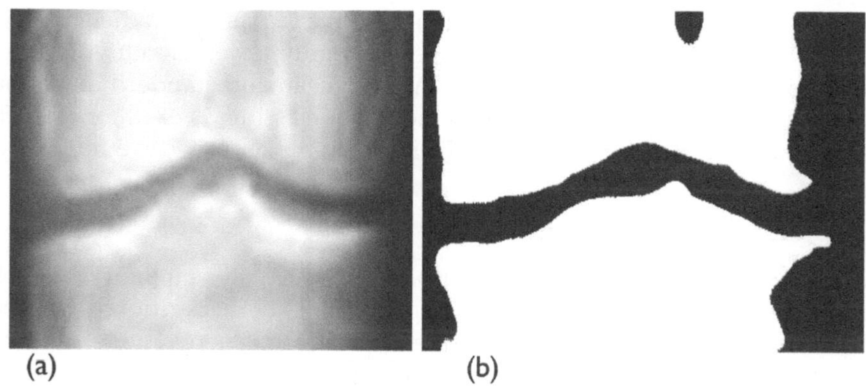

FIGURE 5.4
The result of the segmentation process (a) Original image and (b) Segmented image.

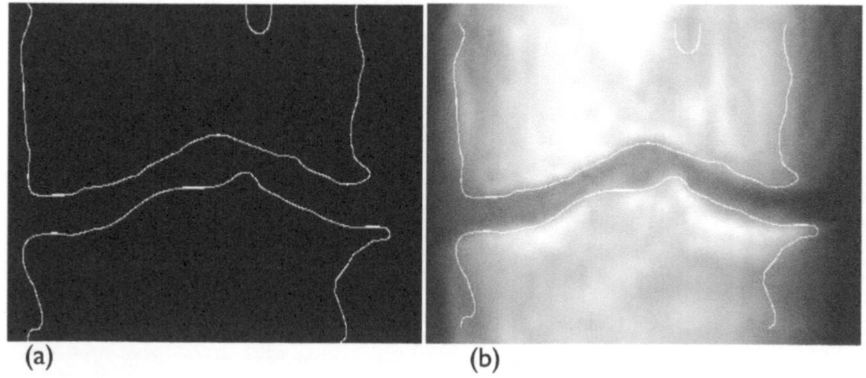

FIGURE 5.5
Boundary extraction (a) Boundary using the morphological technique, (b) Boundary over the synovial cavity region.

5.6.5 Edge Curvature Computation

The primary purpose of extracting boundaries from the segmented images is to locate the boundary of the synovial cavity region. However, there are possibilities of generation of false object boundaries outside the synovial cavity region. Generally, these edges will be smaller in size. Such false contours are eliminated, and larger edges are retained for further processing. The properties of the synovial cavity region will help in the classification of the image as normal or affected by OA. For normal images, the edges will be smooth with less variation in curvature values, whereas for the abnormal images the edges will have high local curvature. Figure 5.6 (a) shows long edges remaining after elimination of false contour, and Figure 5.6 (b) is the curvature profile of these edge segments.

5.6.6 Classification

The mean ($\bar{X}$) and standard deviation (σ) are computed from the array of curvature values by using Equations 5.1 and 5.2 respectively.

$$= \frac{1}{n}\sum_{1}^{n} X_i \text{ and } \sigma = \sqrt{\frac{1}{n}\sum_{i=1}^{n}(x_i -)^2} \qquad (5.1) \text{ and } (5.2)$$

The mean and standard deviation for the curvature profile shown in Figure 5.6 (b) are 0 and 1.26 respectively. Mean and standard deviation values are computed for all the images and a k-nearest neighbors (k-NN) classifier is used to classify the images into two groups: normal and affected with OA. The k-NN classifier is a nonparametric method used for classification [25, 27] as well as for regression. It has two phases: training and classification. The result of the k-NN classifier is a class member. The classification

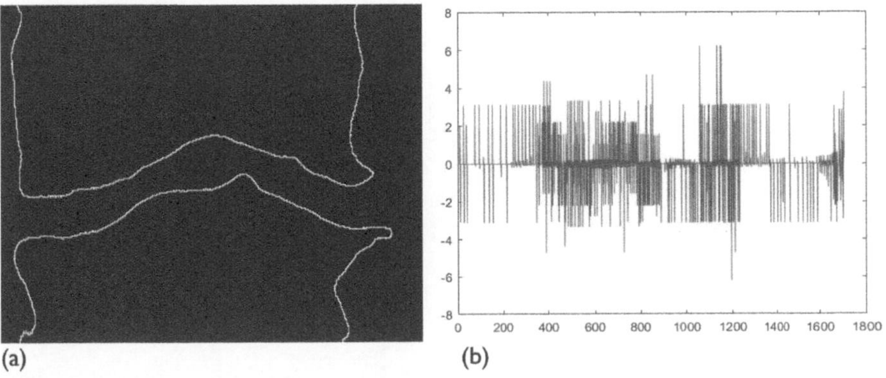

FIGURE 5.6
(a) Large edge segments and (b) Curvature profile of edges.

of an object will depend on the majority vote by its neighboring members, and the object will be assigned to the class in which it is nearest among the k-nearest neighbors.

5.6.7 Results and Discussion

For experimentation, eight normal images and 10 images with a different level of OA were collected from a local hospital. Guidance from radiologists was obtained for identifying whether the images belong to normal patients or for patients suffering from OA. Matlab R2016a software is used for the implementation of the proposed work. k-NNs are used to classify the accuracy of further work with two features, mean and standard deviation.

5.6.7.1 Results for Abnormal Images

Figure 5.7 shows the result of the proposed method on an image of a patient suffering from OA. Figure 5.7 (a) shows the original-ray image. Figure 5.7 (b) and (c) shows the extracted edges and the curvature profile of that edge respectively.

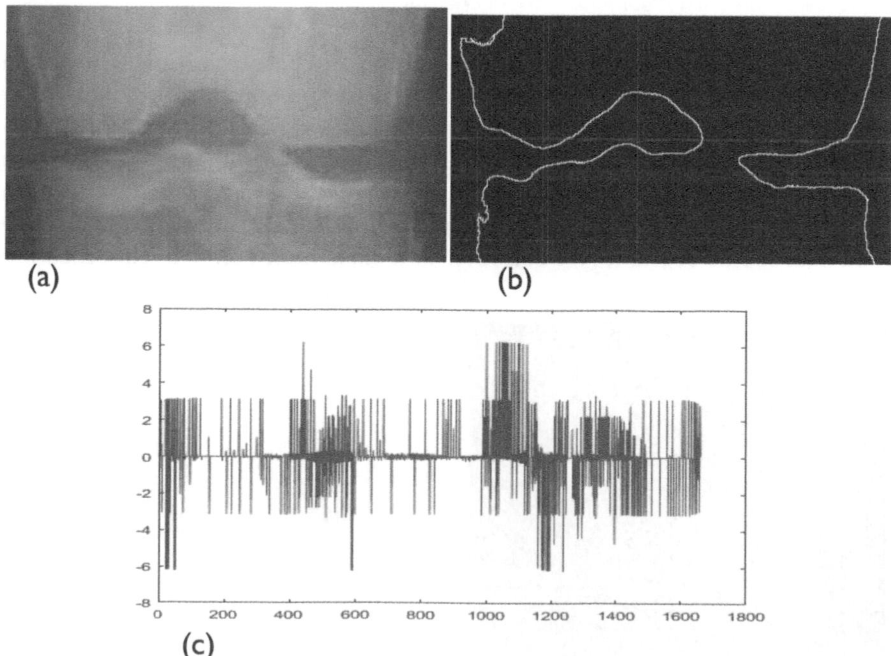

FIGURE 5.7
Result of abnormal image (a) Original image, (b) Large edge segments, and (c) Curvature profile.

5.6.7.2 Results for Normal Images

Figure 5.8 shows the result of the proposed method on an image of a healthy bone. Figure 5.8 (a) shows the original-ray image, whereas Figure 5.8 (b) and (c) shows the extracted edges and the curvature profile of that edge.

5.6.7.3 Classification Results

The classification result of the proposed work (Figure 5.9) has an accuracy of 84%. The scatter diagram shows a plot of two features: mean and standard deviation.

5.7 Texture Analysis-Based Feature Extraction Method

In this method, an adaptive thresholding-based segmentation method is applied to extract the desired portion. To locate the synovial cavity region with precision, the bounding box is drawn to isolate the synovial cavity

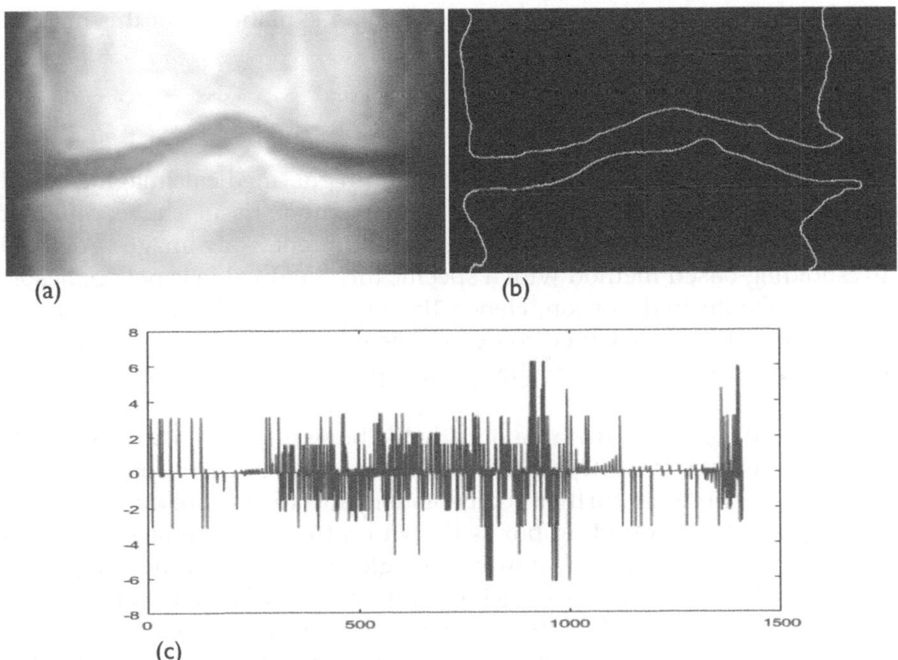

FIGURE 5.8
Result of normal image (a) Original image (b) Large edge segment, and (c) Curvature profile.

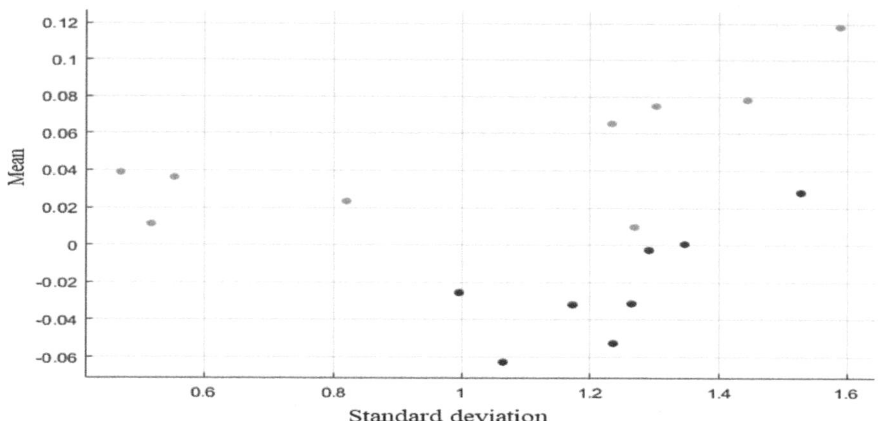

FIGURE 5.9
Scatter plot showing classification results with blue dots representing normal images and orange dots representing images with OA.

region from the background. After that, several texture-based GLCM features such as contrast, correlation, energy, and homogeneity are extracted for classification purposes. SVM and k-NN classifiers are used to categorize the images into the respective classes. The detailed explanation of this method is provided in the upcoming sections.

5.7.1 Segmentation

Intensity variation is a significant problem in the medical imaging-based analysis system [26]. This means that the same bone tissue shows different intensity values to the same bone tissues in different X-ray images. A global thresholding-based method with a specific threshold value is not applicable to extract the desired portion. Hence, the adaptive thresholding-based segmentation method is adapted to extract the synovial cavity region from an image. Figure 5.10 (a) and (b) shows the input X-ray image and segmented image respectively.

Figure 5.10 (b) shows the area outside the bone region is also quite dark. This area belongs to neither bone nor synovial cavity. To avoid unnecessary complications in further segmentation due to the presence of these darker portions, a bounding box is drawn on the binary image as shown in Figure 5.11 (a), which will draw a rectangle on all sides of the image such that every side of the box will touch minimum one white pixel in the binary image. Since the synovial cavity region will also be darker, this step will help in locating the central part of the synovial cavity region. Figure 5.11 (b) is that

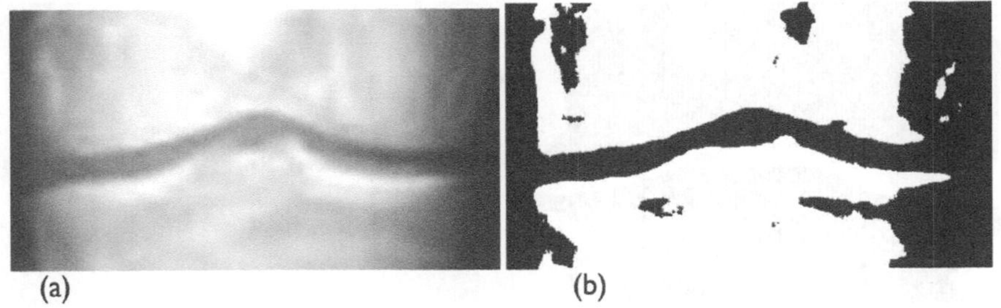

FIGURE 5.10
(a) Original X-ray image and (b) Segmented image.

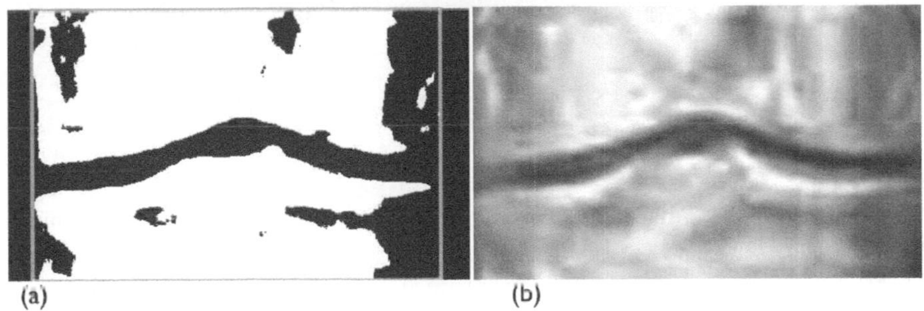

FIGURE 5.11
Cropping operation (a) Image with the bounding box and (b) Cropped image.

portion of the enhanced image which is within the bounding box marked with the blue color shown in Figure 5.11 (a).

5.7.2 Locating the Center of the Synovial Cavity

The histogram property is used to locate the center of the synovial cavity region. In this step, the binary image is first inverted, so that the pixels belonging to the synovial cavity region appear as white, and the bone area pixels appear as black. In the inverted image the maximum number of white pixels is in the rows belonging to the synovial cavity region. The row containing the maximum number of white pixels is located in this image and indicated with a blue line in Figure 5.12 (a). Figure 5.12 (b) shows the same line in the enhanced image. The next task will be to remove the unwanted region from the image by retaining the region around the synovial cavity. This task is performed by empirically selecting 60 rows above and below the center line and extracting that region with the process of feature extraction as shown in Figure 5.12 (c).

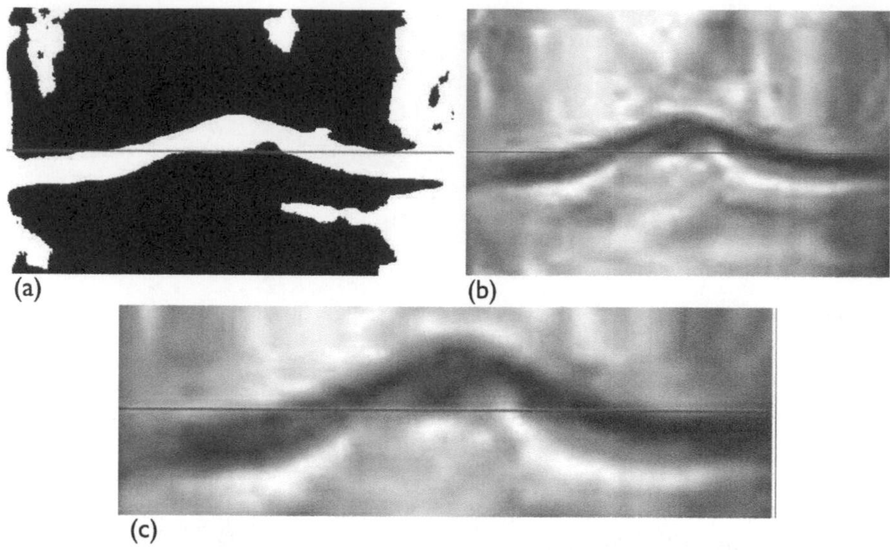

FIGURE 5.12
Location of the center of the synovial cavity region (a) Inverted image, (b) Enhanced image, and (c) Cropped image.

5.7.3 Feature Extraction

GLCM features from the synovial cavity region are extracted for further classification. GLCM, also known as the gray-level spatial dependence matrix, is a statistical procedure to analyze the texture of the image [26, 28], which considers the spatial relationship among image pixels. Based on GLCM, the texture properties of an image may be extracted by calculating how often pairs of the pixel with specific values occur in an image, and what type of spatial relationship exists among pixels may also be established. We extracted four image texture features based on GLCM: contrast, correlation, energy, and homogeneity. The contrast, which is intensity contrast among the entire image pixel and its neighbor, is (0.0690, 0.1181) for the image in Figure 5.12 (c). The correlation is a statistical measure showing how a pixel correlated with its neighbor in the entire image, and it is in the range (0.9876, 0.9788) for the image in Figure 5.12 (c). The energy, which is the sum of the squared elements, is (0.1457, 0.1321), and homogeneity, which measures the closeness of the element distribution in GLCM to the GLCM diagonal, is (0.9655, 0.9410). These are the range values.

5.7.4 Classification

Two classifiers, cubic SVM and k-NN, are used for classification. Since this work is a two-class problem, SVM is one of the ideal classifiers, and since the features are not separable, cubic SVM is chosen here. The SVM algorithm is generally used for pattern classification and regression. During the training, the SVM algorithm finds the optimal linear hyperplane in such a way that the expected classification error for unknown test samples gets minimized. We used k-NN for classification and regression problems; moreover, it is a useful technique that is used to assign weight to the contributions of the neighbors. k-NN is a suitable algorithm to cluster the data according to properties.

5.7.5 Results and Discussion

The experimentation is carried out using Matlab R2016a on a PC with a I7 processor with 8 GB memory. Fourteen normal and 17 abnormal images were used from the OAI. The proposed algorithm failed to detect the center line of the synovial cavity region for three abnormal images.

5.7.5.1 Results for Abnormal Image

Figure 5.13 (a) shows an image for a patient with OA. The result of cropping the non-bone region in binary form is shown in Figure 5.13 (b). The final segmented synovial cavity of this abnormal image is shown in Figure 5.13 (c). GLCM features were extracted from the image shown in Figure 5.13 (c) and the results obtained are (0.0779, 0.0998) for contrast, (0.9725, 0.9652) for correlation, (0.2172, 0.2073) for energy, and (0.9612, 0.9502) for homogeneity. On comparing these results with the results obtained for normal images, it can

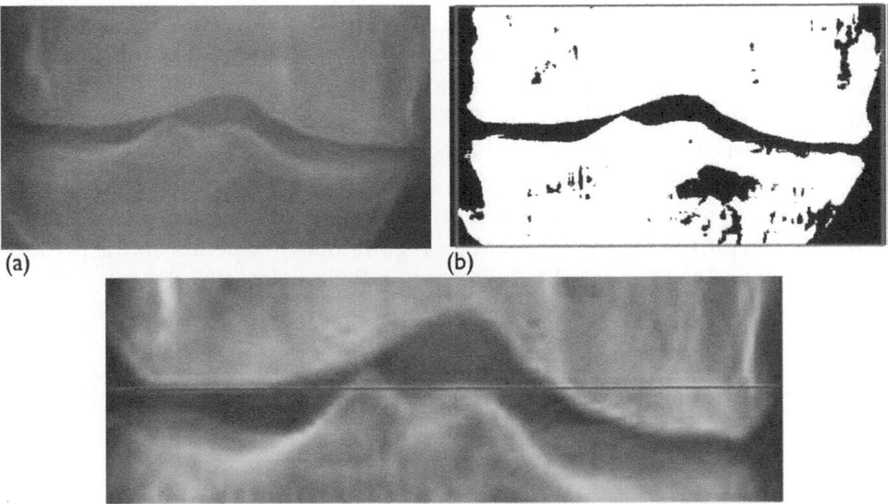

FIGURE 5.13
Result on an image with OA (a) Original image, (b) Bounding box, and (c) Center of synovial cavity region.

be noticed that there is a high variation in the energy feature among these two classes of images. The normal image produced lower energy as compared to the abnormal image.

5.7.5.2 Comparison

Our proposed method could classify both normal and abnormal images correctly using k-NN and cubic SVM classifier. The classification rate for normal images using k-NN classifier is 100%, but for SVM it is 79%. For abnormal images, k-NN and SVM give 100% accuracy. The overall classification accuracy of the SVM classifier is 89%. These two classification methods validate with four-fold cross-validation. Results are compared with the work proposed by GW Stachowiak et al. [23], which has six stages: (1) Selected trabecular bone (TB) texture ROIs. (2) Measurement of distances between TB texture images. (3) Generation of a classifier ensemble. (4) Selection of accurate classifiers. (5) Classification of the ROIs. (6) classified ROIs. For classification purposes, they have used a dissimilarity-based multiple classifier (DMC), which gives 90.51% accuracy with two-fold cross-validation (Table 5.1).

5.7.5.3 Failure Analysis

As stated earlier, the proposed method successfully identified the central part of the synovial cavity region, but it failed to do so for three abnormal images. One of the reasons for its failure is the fact that in cases of images of patients suffering from OA, the gap between the upper and lower bones will be narrow. With the increase in the level of severity, this gap reduces further. In such circumstances, the number of pixels belonging to the synovial cavity region will also come down. Because of this reason, the algorithm will falsely trace the synovial cavity region in some other part of the bones other than in the actual synovial cavity as shown in Figure 5.14. In Figure 5.14(b) it can be noticed that the center part of the synovial cavity region, shown with the blue colored line, is falsely identified at the bottom part of the image. The proposed algorithm also fails if the bone part in the X-ray image is skewed, since the identification of the central part of the synovial cavity region totally depends on the position of the bone and it is expected to be correct in the vertical direction.

TABLE 5.1

Classification Results and Comparison

Image	k-NN	Cubic SVM	DMC-based
Normal	100%	79%	–
Abnormal	100%	100%	–
Overall	100%	89%	90.51

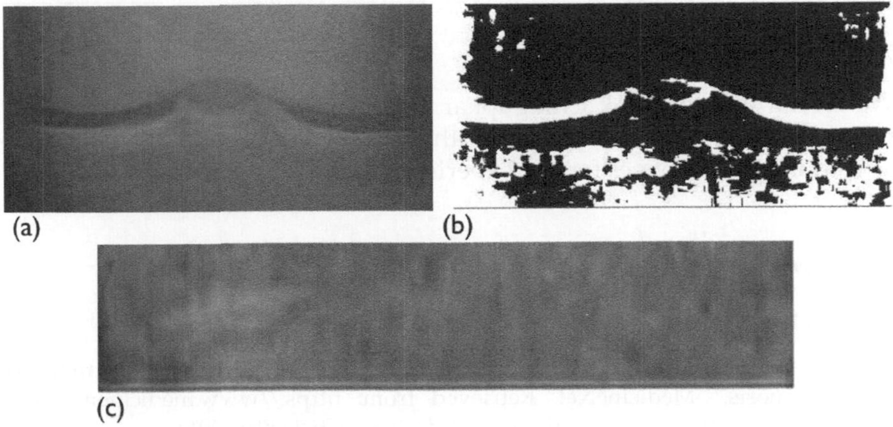

FIGURE 5.14
Failure to locate the central part of the synovial cavity region (a) Abnormal image, (b) False location of the center row, (c) False segmentation.

5.8 Conclusion

In the proposed work on curvature-based and texture-based edges, two different feature extraction methods are used to analyze and classify the images into two classes: normal and abnormal. At first, patient-specific X-ray images are collected from local hospitals in India. Wiener filter and histogram equalization-based image enhancement are used to remove unwanted artifacts and to enhance the desired part respectively. In the edge curvature-based method, the multistage segmentation method, which is a combination of active contour and level set, is used to separate the desired region from the background; moreover, it is also employed to extract the synovial cavity region. After that, edge curvature-based features such as edge length and edge curvature profile are extracted and passed to the SVM classifier to categorize images into their respective classes. In the texture-based method, a simple adaptive thresholding-based segmentation technique is adopted to extract the synovial cavity region. After that, the bounding box is used to locate the center part of the synovial cavity. Then GLCM features such as skewness, kurtosis, standard deviation, and energy are extracted for classification purpose. SVM and k-NN classifiers are used to categorize the image into two classes: normal and abnormal. The k-NN classifier shows 100% accuracy for both normal and OA affected images. The proposed classification systems classify the image into two classes. In the future, we aim to develop a system that will classify image in five grades using deep learning-based classification methods.

Acknowledgment

We would like to thank Chidgupkar Hospital Pvt. Ltd., a multi-specialty hospital in India, for providing us with the X-ray images and giving permission to use these images for our experimentation.

References

1. C. William and J. R. Shiel, "Osteoarthritis (OA): Treatment, symptoms, Diagnosis," MedicineNet. Retrieved from: https://www.medicinenet.com/osteoarthritis_overview_pictures_slideshow/article.htm, 2016.
2. A. Rastogi, "Osteoarthritis," NHP. Retrieved from: https://www.nhp.gov.in/disease/musculo-skeletal-bone-joints-/osteoarthritis, 2017.
3. J. H. Kellgren and J. S. Lawrence, "Radiological assessment of osteoarthrosis," *Annals of the Rheumatic Diseases*, vol. 16, no. 4, p. 494, 1957.
4. J. H. Ryu, A. Lee, M. S. Huh, J. Chu, K. Kim, B. S. Kim, and I. Youn, "Measurement of MMP activity in synovial fluid in cases of osteoarthritis and acute inflammatory conditions of the knee joints using a fluorogenic peptide probe-immobilized diagnostic kit," *Theranostics*, vol. 2, no. 2, p. 198, 2012.
5. N. Gerwin, C. Hops, and A. Lucke, "Intraarticular drug delivery in osteoarthritis," *Advanced Drug Delivery Reviews*, vol. 58, no. 2, pp. 226–242, 2006.
6. D. D. Ruikar, R. S. Hegadi, and K. C. Santosh, "A systematic review on orthopedic simulators for psycho-motor skill and surgical procedure training," *Journal of Medical Systems*, vol. 42, no. 9, p. 168, 2018.
7. D. I. Navale, R. S. Hegadi, and N. Mendgudli, "Block-based texture analysis approach for knee osteoarthritis identification using SVM," in *Electrical and Computer Engineering (WIECON-ECE), 2015 IEEE International WIE Conference on*, 2015, pp. 338–341, IEEE.
8. J. Duryea, T. Iranpour-Boroujeni, J. E. Collins, C. Vanwynngaarden, A. Guermazi, J. N. Katz, and C. Ratzlaff, "Local area cartilage segmentation: a semiautomated novel method of measuring cartilage loss in knee osteoarthritis," *Arthritis Care & Research*, vol. 66, no. 10, pp. 1560–1565, 2014.
9. L. Anifah, I. K. E. Purnama, M. Hariadi, and M. H. Purnomo, "Osteoarthritis classification using self-organizing map based on Gabor kernel and contrast-limited adaptive histogram equalization," *The Open Biomedical Engineering Journal*, vol. 7, p. 18, 2013.
10. J. Antony, K. McGuinness, N. E. O'Connor, and K. Moran, "Quantifying radiographic knee osteoarthritis severity using deep convolutional neural networks," in *Pattern Recognition (ICPR),* 2016 *23rd International Conference on*, 2016, pp. 1195–1200, IEEE.
11. L. Anifah, M. H. Purnomo, T. L. R. Mengko, and I. K. E. Purnama, "Osteoarthritis severity determination using self organizing map based gabor kernel," in *IOP Conference Series: Materials Science and Engineering* (Vol. 306, No. 1), 2018, p. 012071, IOP Publishing.

12. P. P. Kawathekar and K. J. Karande, "Use of textural and statistical features for analyzing severity of radio-graphic osteoarthritis of knee joint," in *Information Processing (ICIP), 2015 International Conference on*, 2015, pp. 1–4, IEEE.
13. H. Oka, S. Muraki, T. Akune, A. Mabuchi, T. Suzuki, H. Yoshida, S. Yamamoto, K. Nakamura, N. Yoshimura, and H. Kawaguchi, "Fully automatic quantification of knee osteoarthritis severity on plain radiographs," *Osteoarthritis and Cartilage*, vol. 16, no. 11, pp. 1300–1306, 2008.
14. R. S. Hegadi and D. I. Navale, "Quantification of synovial cavity from knee X-ray images," in 2017 *International Conference on Energy, Communication, Data Analytics and Soft Computing (ICECDS)*, 2017, pp. 1688–1691. IEEE.
15. J. Fripp, S. Crozier, S. K. Warfield, and S. Ourselin, "Automatic segmentation of the bone and extraction of the bone–cartilage interface from magnetic resonance images of the knee," *Physics in Medicine & Biology*, vol 52, no. 6, p. 1617, 2007.
16. F. K. Patel and M. Singh, "Segmentation of cartilage from knee MRI images using the watershed algorithm," *International Journal of Advance Research, Ideas and Innovations in Technology*, vol. 4, no. 2, pp. 1727–1730, 2018.
17. A. Tiulpin, J. Thevenot, E. Rahtu, P. Lehenkari, and S. Saarakkala, "Automatic knee osteoarthritis diagnosis from plain radiographs: a deep learning-based approach," *Scientific Reports*, vol. 8, no. 1, p. 1727, 2018.
18. M. S. Pandey, B. Rajitha, and S. Agarwal, "Computer assisted automated detection of knee osteoarthritis using x-ray images," *Science & Technology*, vol. 1, no. 2, pp. 74–79, 2015.
19. R. Riad, R. Jennane, A. Brahim, T. Janvier, H. Toumi, and E. Lespessailles, "Texture analysis using complex wavelet decomposition for knee osteoarthritis detection: data from the osteoarthritis initiative," *Computers & Electrical Engineering*, vol. 68, pp. 181–191, 2018.
20. R. G. Brown and P. Y. Hwang, *Introduction to Random Signals and Applied Kalman Filtering* (Vol. 3), Wiley, New York, 1992.
21. A. M. Reza, "Realization of the contrast limited adaptive histogram equalization (CLAHE) for real-time image enhancement," *Journal of VLSI Signal Processing Systems for Signal, Image and Video Technology*, vol. 38, no. 1, pp. 35–44, 2004.
22. T. F. Chan and L. A. Vese, "Active contours without edges," *IEEE Transactions on Image Processing*, vol. 10, no. 2, pp. 266–277, 2001.
23. G. W. Stachowiak, M. Wolski, T. Woloszynski, and P. Podsiadlo, "Detection and prediction of osteoarthritis in knee and hand joints based on the X-ray image analysis," *Biosurface and Biotribology*, vol. 2, no. 4, pp. 162–172, 2016.
24. D. D. Ruikar, R. S. Hegadi, and K. C. Santosh, *Contrast Stretching-Based Unwanted Artifacts Removal from CT Images in Recent Trends in Image Processing and Pattern Recognition* (accepted), 2019, Springer.
25. S. Vajda and K. C. Santosh, "A fast k-nearest neighbor classifier using unsupervised clustering," in *International Conference on Recent Trends in Image Processing and Pattern Recognition*, 2016, pp. 185–193, Springer, Singapore.
26. A. Karargyris, J. Siegelman, D. Tzortzis, S. Jaeger, S. Candemir, Z. Xue, K. C. Santosh, S. Vajda, S. Antani, L. Folio, and G. R. Thoma, "Combination of texture and shape features to detect pulmonary abnormalities in digital chest X-rays," *International Journal of Computer Assisted Radiology and Surgery*, vol. 11, no. 1, pp. 99–106, 2016.

27. K. C. Santosh, L. Wendling, S. Antani, and G. R. Thoma, "Overlaid arrow detection for labeling regions of interest in biomedical images," *IEEE Intelligent Systems*, vol. 31, no. 3, pp. 66–75, 2016.
28. K. C. Santosh, S. Candemir, S. Jäger, L. Folio, A. Karargyris, S. Antani, and G. Thoma, "Rotation detection in chest radiographs based on generalized line histogram of rib-orientations," in *Computer-Based* Medical Systems (CBMS), 2014 IEEE *27th International Symposium on*, 2014, pp. 138–142, IEEE.
29. R. S. Hegadi, U. P. Chavan, and D. I. Navale, "Identification of knee osteoarthritis using texture analysis," in *Data Analytics and Learning*, 2019, pp. 121–129, Springer, Singapore.
30. R. S. Hegadi, T. D. Pawar, and D. I. Navale, "Classification of osteoarthritis-affected images based on edge curvature analysis," in *Data Analytics and Learning*, 2019, pp. 111–119, Springer, Singapore.
31. R. S. Hegadi, T. D. Pawar, D. I. Navale, and D. D. Ruikar, *Osteoarthritis Detection and Classification From Knee X-ray Images Based on Artificial Neural Network in Recent Trends in Image Processing and Pattern Recognition* (accepted), 2019, Springer.

6

Detection and Classification of Non-Proliferative Diabetic Retinopathy Lesions

Ramesh R. Manza, Bharti W. Gawali, Pravin Yannawar, and K.C. Santosh

CONTENTS

6.1 Introduction

Artificial neural networks (ANNs) are computational models inspired by the central nervous system, which is capable of solving complex machine learning as well as pattern recognition tasks. ANNs are usually obtainable as systems of interconnected "neurons" which can compute values from inputs. A typical neural network for diabetic retinopathy lesions recognition is defined by a set of input neurons activated by the pixels of an input fundus image. The activation of these neurons is processed on weight and transformed by a function determined by the network's designer, to other neurons. This process is repeated until the final output neuron is activated. It determines which lesions were present on that fundus image. Like other machine learning methods, this system's information about data and neural networks have been used to solve a wide variety of tasks which are hard to solve using ordinary rule-based programming, including computer vision and diabetic retinopathy lesions recognition.

6.2 Methodology

6.2.1 Preprocessing

Preprocessing is a very important part because patient's movement, poor focus, bad positioning, reflections, and inadequate illumination can cause a significant proportion of images to be of such poor quality as to interfere with analysis. In the retinal images there can be variations caused by several factors, including differences in cameras, illumination, acquisition angle, and retinal pigmentation. To see whether a fundus image is normal or abnormal we extracted the mask of it and then we processed that image.

6.2.1.1 RGB Color Separation

In medical image processing, the green channel is widely adopted because it shows more features as compared to the red and blue channels respectively. Below are the formulas for red, green, and blue color separation:

Red channel

$$r = \frac{R}{(R+G+B)} \tag{6.1}$$

Here "r" is a red channel and R, G, and B are red, green, and blue respectively.
Green channel

$$g = \frac{G}{(R+G+B)} \tag{6.2}$$

Here "g" is a green channel and R, G, and B are red, green, and blue respectively.
Blue channel

$$b = \frac{B}{(R+G+B)} \tag{6.3}$$

Here "b" is a blue channel and R, G, and B are red, green and blue respectively.

6.2.1.2 Mask Separation

The fundus can be easily separated from the background by extracting the red channel from the RGB fundus image and then applying a threshold on that red channeled image for the extraction of the fundus mask.
Threshold

$$\mathrm{T} = \frac{1}{2}(m_1 + m_2) \tag{6.4}$$

Here m_1 and m_2 are the intensity values.

6.2.1.3 Image Enhancement

Before image enhancement, we extract the green channel from the RGB fundus image after the extraction of the green channel performs histogram equalization.

6.2.1.4 Histogram Equalization

Let $p_s(s)$ and $p_d(d)$ represent the standard image and desired image probability density functions, respectively. The histogram equalization of the standard image is as follows:

$$u = T(s) = \int_0^s p_s(x)\,dx \tag{6.5}$$

The histogram equalization of the desired image is obtained by a similar transformation function as follows:

$$v = Q(d) = \int_0^d p_d(x)\,dx \tag{6.6}$$

The values of d for the desired image are obtained as follows:

$$d = Q^{-1}[u] = Q^{-1}[T(s)] \tag{6.7}$$

A standard retinal image is used as a reference for the histogram specification technique in agreement with the expert ophthalmologist.

6.2.2 Removal of Optic Disc from Fundus Images

The optic disc (OD) or optic nerve head (ONH) is the place where ganglion cell axons exit the eye to form the optic nerve. There are no light-sensitive rods or cones to respond to a light stimulus at this point. At the time of detection of non-proliferative lesions, the ODand the exudates (EXs) have same color, and sometimes also the same shape and intensity values. This makes it very hard for the system to recognize the OD and EXs. Sometimes, even experts can get confused when trying to recognize the OD and EXs. For the removal of OD we use enhanced image by histogram equalization and then apply a complement function on a histogram-equalized image. Once the complement is performed, the intensity transformation function is applied. After intensity transformation, we perform a subtraction operation: we subtract the complemented image from the intensity-transformed image. This subtraction gives the output as an OD-free image.

The complement function is performed as follows:

$$A^c = \{\omega \mid \omega \notin A\} \tag{6.8}$$

Here A^c is a complement, ω is the element of A, ∉ stands for not an element of A, and A is set.

Similarly, the intensity transformation function can be computed as

$$s = T(r), \tag{6.9}$$

where T is transformation and r is intensity

To remove the OD, we use green channel image; after extraction of green channel, we perform histogram equalization for enhancement, and after enhancement use complement function for highlighting the OD. The intensity transformation function is utilized for contrast adjustment. After these operations, subtraction operations were performed to remove the OD.

6.3 Detection of Microaneurysms

Ophthalmological clinical examination reveals that microaneurysms (MAs) are small circular deep-red dots observed in the fundus. They are not visible to the necked eye. In order to identify the microaneurysms, doctors recommend the fluorescein angiography technique. During fluorescein angiography, a dye is injected into a vein in your arm. Once injected, it takes about 10 to 15 seconds to circulate through your body. As the dye enters the blood vessels in your eyes, a series of photos are taken to chart the dye's progress. More pictures are taken after most of the dye has passed through your eyes to see if any of it has leaked out of the blood vessels or highlight the fine details of fundus. However, retinal angiography is not recommended for children, pregnant women, or patients suffering for cardiovascular disorders. Moreover, there is a chance of infection if the skin is broken. In rare cases, if a person is hypersensitive to the dye, they may experience dizziness or faintness, dry mouth or increased salivation, hives, increased heartrate, a metallic taste in their mouth, nausea and vomiting, sneezing.

6.4 Detection of Hemorrhages

Hemorrhages are one of the first signs of diabetic retinopathy and are also prominent in other ocular diseases. They can manifest in different forms: small round-dot hemorrhages that are related to MAs and indistinguishable from them in color fundus images; flame-shaped and blotch (cluster) hemorrhages whose names describe their appearance; or larger boat-shaped hemorrhages. For the detection of retinal hemorrhages, a high-resolution fundus image is taken, then the green channel is extracted from the RGB image. Once it is extracted, contrast limited adaptive histogram equalization (CLAHE) is performed for enhancement of hemorrhages. To extract the hemorrhages, symlet wavelet (level 1) is performed in order to find the exact location of hemorrhages. CLAHE is performed by:

$$g = \left[g_{max} - g_{min}\right] * p(f) + g_{min}, \tag{6.10}$$

where

g_{max} = maximum pixel value
g_{min} = minimum pixel value
g = computed pixel value
$P(f)$ = CPD (cumulative probability distribution)

For exponential distribution, gray level can be adapted as

$$g = g_{min} - \left(\frac{1}{\alpha}\right) * \ln\left[1 - p(f)\right], \tag{6.11}$$

where

α = clip parameter

The formulas for green channel separation, threshold, and symlet are explained in Sections 6.3 and 6.4.

6.5 Detection of EXs

EXs are one of the most commonly occurring lesions. They are associated with patches of vascular damage with leakage. The size and spreading of EXs may vary during the progress of the disease. There are tiny yellow patches of hard EXs which are fats from the blood on the retina. EXs are marked by masses of white or yellowish EX in the posterior part of the fundus oculi, with deposits of cholesterin. They are one of the primary signs of diabetic retinopathy, which is a main cause of blindness and could be prevented with an early screening process. Pupil dilation is required in the normal screening process but this affects the patient's vision.

6.6 Extraction of Retinal Blood Vessels

The retina is the light-sensitive tissue that lines the inside of the eye. The retina functions in a manner similar to film in a camera. The optical elements within the eye focus an image onto the retina, initiating a series of chemical and electrical events within it. Nerve fibers in the retina send electrical signals to the brain, which then interprets these signals as visual images. There are two circulations to the retina, both supplied by the ophthalmic artery, the first branch of the internal carotid artery on each side. The outer and middle retinal layers, including the outer plexiform and outer nuclear layers, the photoreceptors, and the retinal pigment epithelium, are nourished by branches of the posterior ciliary arteries, which enter the back of the eye outside the optic nerve. Retinal neovascularization lacks the bifurcating pattern of normal vessels. Neovascularization causes blindness because all the retina gets nourished with the blood vessels. Retinal blood vessels extraction is very important because after it is performed we can calculate the diameter of vessels to see whether they are normal or not. The normal diameter of overall blood vessels is >25 mm to 30 mm. Figure 6.1 shows the extraction of retinal blood vessels.

For the extraction of retinal blood vessels, high-resolution fundus images are taken, the green channel of color image is extracted because green channels shows the intensity of the image as compared to red and blue respectively. After the green channel extraction, the intensity transformation function is perfomed in order to enhance the fundus image, then histogram equalization is applied on the intensity-transformed image to highlight the retinal blood vessels. After this, a morphological open function is performed on a histogram-equalized image to thin the blood vessels. But when we applied the thinning operation, some salt-and-pepper kind of noise got added. To remove that noise we used the median filter, and to extract blood vessels we performed a threshold operation. After the threshold we got the extracted blood vessels, but the exact vessels network was not showing; therefore, symlet wavelet

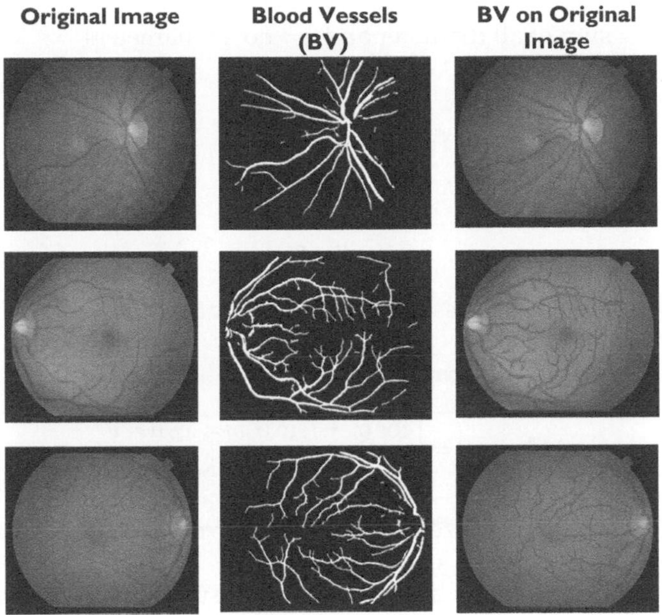

FIGURE 6.1
Extraction of retinal blood vessels.

needs to be applied with level 1 because only at this level do we get the exact blood vessels. When we increase the level of symlet wavelet, the vessels get thinner. That's why we again go back with level 1 only. After the extraction of retinal blood vessels we calculate the area, diameter, length, thickness, mean diameter, tortuosity, venous beading, and bifurcation points of the extracted blood vessels. Below are the formulas for the area, diameter, length, thickness, mean diameter, tortuosity, venous beading, and bifurcation points.

Area

$$Area = \pi \times r^2 \tag{6.12}$$

Diameter

$$Diameter = \sqrt{Area / \pi} \tag{6.13}$$

Length

$$Length = \frac{Area}{2} \tag{6.14}$$

Thickness

$$Thickness = \frac{Area}{Length} \tag{6.15}$$

Mean Diameter

$$Mean = \frac{\sum X}{n} \tag{6.16}$$

where $\sum X$ = sum of all diameter and n = no. of diameters

Tortuosity

$$Tortuosity = \frac{Length}{Distance} \tag{6.17}$$

Venous Beading

$$Area = 5 * Lenght \tag{6.18}$$

If the area of the blood vessels equals 5 * length, then venous beading is present.

Minutia technique (bifurcation points)

$$\mathcal{M}^{m}(m,n) = \begin{cases} \cos\alpha_i + j\sin\alpha_j m = x_i, n = y_i \\ 0 \; Otherwise \end{cases}, \tag{6.19}$$

where $(x_i, y_{i,}\alpha_i) \in M$is the size of the *minutiae direction map* (MDP) and $\mathcal{M}^{m}$ is the image.

All the parameter details are present in the subsequent sections (experimental results). After detecting the non-proliferative diabetic retinopathy lesions (NPDRs), apply ANN for grading of lesions. An ANN refers to the interconnections between the neurons in the different layers of each system. An example system has three layers. The first layer has input neurons that send data via synapses to the second layer of neurons, and then via more synapses to the third layer of output neurons. The synapses store parameters called "weights," which manipulate the data in the calculations.

After grading a NPDR, apply the receiver operating characteristic curve. In this performance analysis, we calculate TP, FP, FN, and TN; after that, we calculate the TP rate, FP rate, precision, recall, and F-measure. Following are the formulas for the parameters of the ROC curve.

TP rate:

$$\text{TP Rate} = \frac{TP}{(TP + FN)} \tag{6.20}$$

FP rate:

$$\text{FP Rate} = \frac{FP}{(TN + FP)} \tag{6.21}$$

Precision:

$$\text{Precision} = \frac{TP}{(TP + FP)} \tag{6.22}$$

Recall:

$$\text{Recall} = \frac{TP}{(TP + FN)} \tag{6.23}$$

F-measure:

$$\text{F-measure} = \frac{2 * TP}{(2 * TP + FP + FN)} \tag{6.24}$$

6.7 Experimental Work

NPDR is an early stage of diabetic retinopathy. In NPDR, tiny blood vessels within the retina leak blood or fluid and this leakage causes the damage to the retina. NPDR is categorized in three stages, namely, mild, moderate, and severe NPDR. For the detection of NPDR lesions such as MAs, EXs, hemorrhages, and neovascularization, we perform digital image processing techniques such as histogram equalization, intensity transformation function, and morphological operations, etc., and for extracting the feature symlet wavelet, level 1 is used. Following are the experimental setups.

- Extraction of mask
- Removal of OD
- Detection of MAs
- Detection of EXs
- Detection of hemorrhages
- Extraction of retinal blood vessels

When preprocessing, we start with the green channel separation. Here we extract the green channel from the RGB image, because the green channel shows a high-intensity image as compared to red and blue respectively.

6.7.1 Extraction of Mask

The mask separation from the fundus image is a very important part because, when processing the images for lesion detection and extraction, we have to process only good-quality images. Here we extract the mask of the fundus images to get the exact shape of the retina. If that shape is not good, then we ignore that image for the further processing. In Figure 6.2, some images of DiarectDB0 (fundus image database) have a corrupted mask. To avoid these kinds of images, we separate the mask from the fundus images and remove the corrupted images for further processing. Figure 6.2 shows the extracted mask of normal and abnormal fundus images.

6.7.2 Removal of OD

The OD is mostly the brighter part of the retina. The OD or ONH is the place where ganglion cell axons exit the eye to form the optic nerve (Figure 6.3).

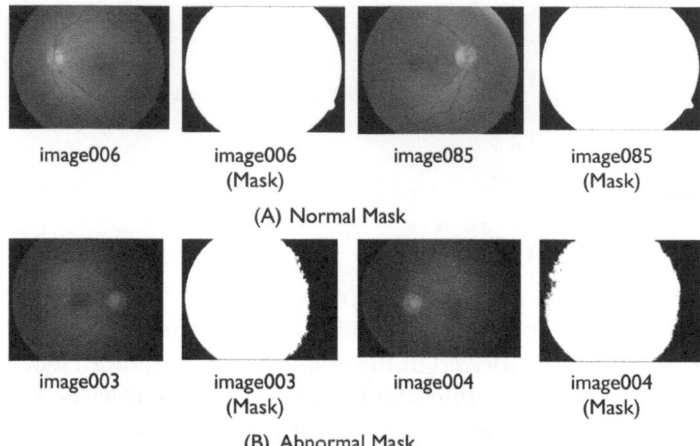

FIGURE 6.2
DiarectDB0 database (A) Normal mask, (B) Abnormal mask.

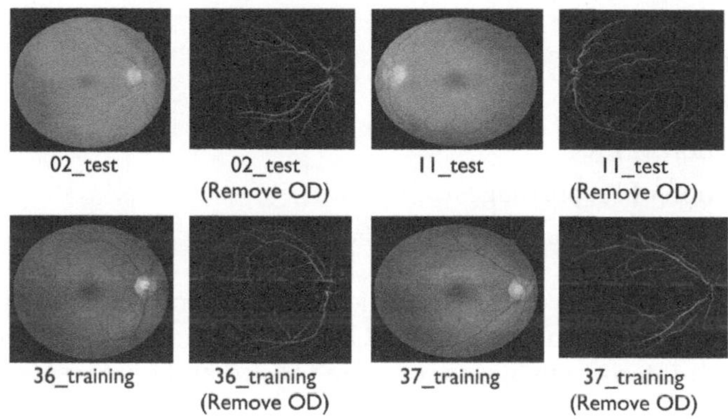

FIGURE 6.3
OD removal.

OD removal is essential because, at the time of detection of non-proliferative lesions, OD and EXs have the same color, and sometimes also the same shape and intensity values. This makes it very hard for the system to recognize OD and EXs. To overcome this complexity, we design one algorithm to remove the OD from the fundus images.

6.7.3 Detection of MAs

MAs are the first clinically detected lesions. They are tiny swellings in the wall of a blood vessel. They appear in the retinal capillaries as small, round, red spots, and are located in the inner nuclear layer of the retina. Figure 6.4 shows detected MAs.

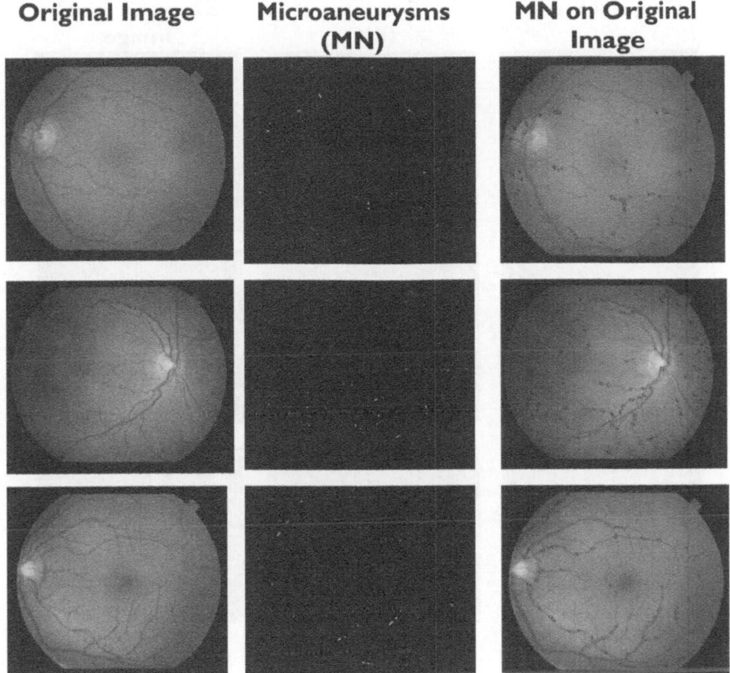

FIGURE 6.4
Detection of MAs.

6.7.4 Detection of EXs

EXs are caused by the breakdown of the blood–retina barrier, and allow leakage of serum proteins, lipids, and protein from the vessels. Figure 6.5 shows some extracted EXs.

6.7.5 Detection of Hemorrhages

Hemorrhages are located in the middle layer of the retina. A retinal hemorrhage is the abnormal bleeding of the blood vessels in the retina. They have a "dot" and "blot" configuration. Figure 6.6 shows the detection of hemorrhages.

6.7.6 Statistical Techniques on NPDR Lesions

Statistical techniques are applied on NPDR lesions in the language of mean, standard deviation, variance, and correlation coefficient. The coefficient (r), can take a range of values from +1 to –1. A value of 0 indicates that there is no association between the two variables. A value greater than 0 indicates a positive association; that is, as the value of one variable increases, so does the

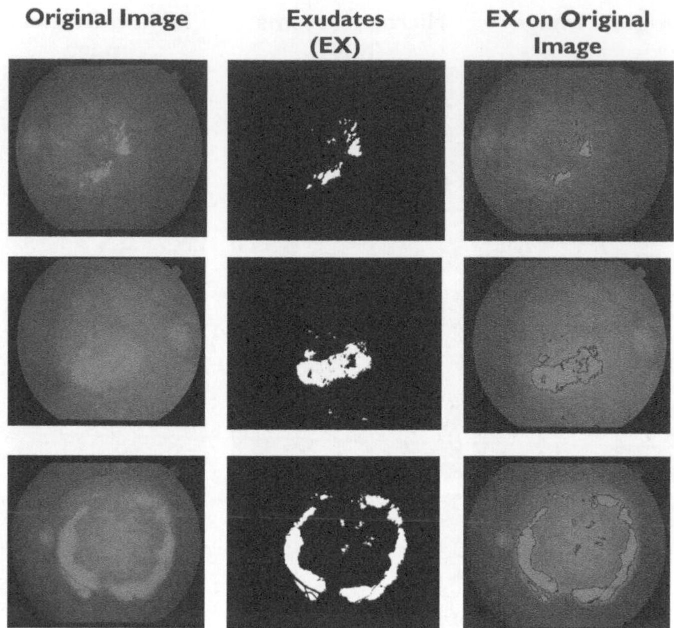

FIGURE 6.5
Detection of EXs.

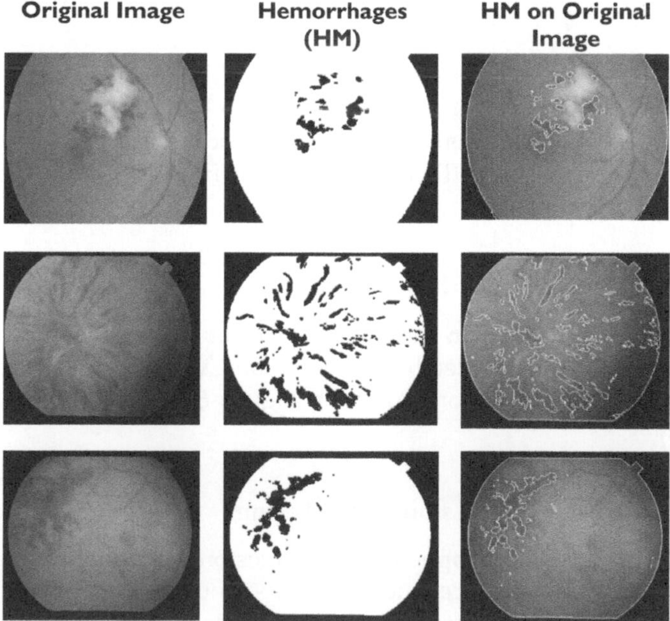

FIGURE 6.6
Detection of hemorrhages.

value of the other variable. A value lower than 0 indicates a negative association; that is, as the value of one variable increases, the value of the other variable decreases.

6.7.6.1 Statistical Techniques on MAs

Table 6.1

TABLE 6.1

Table for MAs

Sr. No.	Image	(x)	(y)	$(x-\bar{X})$	$(y-\bar{Y})$	xy
01	image002	612	912	595.61	895.24	558144
02	image003	984	985	967.61	968.24	969240
03	image004	932	932	915.61	915.24	868624
04	image005	889	905	872.61	888.24	804545
05	image006	204	204	187.61	187.24	41616
06	image007	795	795	778.61	778.24	632025
07	image008	891	891	874.61	874.24	793881
08	image009	138	138	121.61	121.24	19044
09	image0010	137	147	120.61	130.24	20139
10	image0011	688	688	671.61	671.24	473344
11	image0012	474	474	457.61	457.24	224676
12	image0013	100	100	83.61	83.24	10000
13	image0014	136	136	119.61	119.24	18496
14	image0015	143	143	126.61	126.24	20449
15	image0016	149	149	132.61	132.24	22201
16	image0017	31	31	14.61	14.24	961
17	image0018	88	89	71.61	72.24	7832
18	image0019	320	322	303.61	305.24	103040
19	image0020	734	734	717.61	717.24	538756
20	image0021	817	817	800.61	800.24	667489
21	image0022	238	238	221.61	221.24	56644
22	image0023	676	676	659.61	659.24	456976
23	image0024	276	276	259.61	259.24	76176
24	image0025	680	680	663.61	663.24	462400
25	image0026	534	534	517.61	517.24	285156
26	image0027	243	243	226.61	226.24	59049
27	image0028	335	335	318.61	318.24	112225
28	image0029	785	785	768.61	768.24	616225
29	image0030	820	820	803.61	803.24	672400
30	image0031	902	902	885.61	885.24	813604

where

$(x) = Manual, (y) By\ System$

$$\text{Mean } (x) = \frac{491.7}{30} = 16.39$$

$$\text{Mean } (y) = \frac{502.7}{30} = 16.76$$

$$\text{Variance } (x) = \frac{\sum (x - \bar{X})}{N} = \frac{475.31}{30} = 15.85$$

$$\text{Variance } (y) = \frac{\sum (y - \bar{Y})}{N} = \frac{485.94}{30} = 16.19$$

Standard Deviation (x): $\sqrt{Variance(x)} = \sqrt{15.85} = 3.99$

Standard Deviation (y): $\sqrt{Variance(y)} = \sqrt{16.19} = 4.03$

Correlation:

$$r = \frac{\sum (x - \bar{X}) \sum (y - \bar{Y})}{\sqrt{\sum (x - \bar{X})^2 \sum (y - \bar{Y})^2}}$$

where

$\sum (x - \bar{X}) = 475.31,$

$\sum (y - \bar{Y}) = 485.94$

$\sum (x - \bar{X})^2 = 225919.60,$

$\sum (y - \bar{Y})^2 = 236137.69$

$$r = \frac{475.31 * 485.94}{\sqrt{225919.60 * 236137.69}}$$

$$r = \frac{230972.15}{230971.51} = 1$$

The above correlation for the MAs is the positive correlation.

6.7.7 Statistical Techniques on EXs

Table 6.2

TABLE 6.2

Table for EXs

Sr. No.	Image	(x)	(y)	$(x-\bar{X})$	$(y-\bar{Y})$	xy
01	image0032	35	42	33.24	40.19	1470
02	image0033	47	49	45.24	47.19	2303
03	image0034	102	106	100.24	104.19	10812
04	image0035	31	31	29.24	29.19	961
05	image0036	68	68	66.24	66.19	4624
06	image0037	18	18	16.24	16.19	324
07	image0038	41	41	39.24	39.19	1681
08	image0039	96	96	94.24	94.19	9216
09	image0040	89	89	87.24	87.19	7921
10	image0041	48	48	46.24	46.19	2304
11	image0042	21	21	19.24	19.19	441
12	image0043	48	48	46.24	46.19	2304
13	image0044	302	309	300.24	307.19	93318
14	image0045	37	37	35.24	35.19	1369
15	image0046	66	66	64.24	64.19	4356
16	image0047	65	65	63.24	63.19	4225
17	image0048	39	39	37.24	37.19	1521
18	image0049	154	155	152.24	153.19	23870
19	image0050	8	8	6.24	6.19	64
20	image0051	37	37	35.24	35.19	1369
21	image0052	10	15	8.24	13.19	150
22	image0053	13	13	11.24	11.19	169
23	image0054	13	17	11.24	15.19	221
24	image0055	5	5	3.24	3.19	25
25	image0056	43	43	41.24	41.19	1849
26	image0057	51	51	49.24	49.19	2601
27	image0058	5	9	3.24	7.19	45
28	image0059	31	31	29.24	29.19	961
29	image0060	41	41	39.24	39.19	1681
30	image0061	22	28	20.24	26.19	616

where

$(x) = Manual, (y) By\ System$

$$\text{Mean}(x) = \frac{52.86666667}{30} = 1.77$$

$$\text{Mean}(y) = \frac{54.2}{30} = 1.81$$

$$\text{Variance}(x) = \frac{\sum(x-\bar{X})}{N} = \frac{51.11}{30} = 1.71$$

$$\text{Variance}(y) = \frac{\sum(y-\bar{Y})}{N} = \frac{52.40}{30} = 1.75$$

Standard Deviation (x): $\sqrt{Variance(x)} = \sqrt{1.71} = 1.31$

Standard Deviation (y): $\sqrt{Variance(y)} = \sqrt{1.75} = 1.33$

Correlation:

$$r = \frac{\sum(x-\bar{X})\sum(y-\bar{Y})}{\sqrt{\sum(x-\bar{X})^2\sum(y-\bar{Y})^2}}$$

where

$\sum(x-\bar{X}) = 51.11,$

$\sum(y-\bar{Y}) = 52.40,$

$\sum(x-\bar{X})^2 = 2612.24,$

$\sum(y-\bar{Y})^2 = 2745.76$

$$r = \frac{51.11 * 52.40}{\sqrt{2612.24 * 2745.76}}$$

$$r = \frac{2678.17}{2678.17} = 1$$

The above correlation for the EXs is the positive correlation.

6.7.8 Statistical Techniques on Hemorrhages

Table 6.3

TABLE 6.3

Table for Hemorrhages

Sr. No.	Image Name	(x)	(y)	$(x-\bar{X})$	$(y-\bar{Y})$	xy
01	image0062	120070	120080	116399	116409	14418005600
02	image0063	110070	110071	106399	106400	12115514970
03	image0064	131600	131602	127929	127931	17318823200
04	image0065	145410	145414	141739	141743	21144649740
05	image0066	138320	138320	134649	134649	19132422400
06	image0067	98868	98869	95197	95198	9774980292
07	image0068	80622	80627	76951	76956	6500309994
08	image0069	90982	90989	87311	87318	8278361198
09	image0070	113110	113110	109439	109439	12793872100
10	image0071	104120	104120	100449	100449	10840974400
11	image0072	98650	98650	94979	94979	9731822500
12	image0073	82427	82428	78756	78757	6794292756
13	image0074	121290	121290	117619	117619	14711264100
14	image0075	116040	116040	112369	112369	13465281600
15	image0076	80753	80753	77082	77082	6521047009
16	image0077	75480	75480	71809	71809	5697230400
17	image0078	87198	87198	83527	83527	7603491204
18	image0079	193540	193540	189869	189869	37457731600
19	image0080	109350	109350	105679	105679	11957422500
20	image0081	128770	128770	125099	125099	16581712900
21	image0082	121640	121640	117969	117969	14796289600
22	image0083	183450	183450	179779	179779	33653902500
23	image0084	75502	75502	71831	71831	5700552004
24	image0085	75537	75539	71866	71868	5705989443
25	image0086	95540	95540	91869	91869	9127891600
26	image0087	108640	108640	104969	104969	11802649600
27	image0088	94552	94558	90881	90887	8940648016
28	image0089	102800	102800	99129	99129	10567840000
29	image0090	105220	105220	101549	101549	11071248400
30	image0091	113610	113613	109939	109942	12907572930

where

$$(x) = Manual, (y) By\ System$$

$$\text{Mean}(x) = \frac{110105.37}{30} = 3670.17$$

$$\text{Mean}(y) = \frac{110106.77}{30} = 3670.2$$

$$\text{Variance}(x) = \frac{\sum(x-\bar{X})}{N} = \frac{106435.2}{30} = 3547.84$$

$$\text{Variance}(y) = \frac{\sum(y-\bar{Y})}{N} = \frac{106436.57}{30} = 3547.89$$

Standard Deviation (x): $\sqrt{Variance(x)} = \sqrt{3547.84} = 59.57$

Standard Deviation (y): $\sqrt{Variance(y)} = \sqrt{3547.89} = 59.57$

Correlation:

$$r = \frac{\sum(x-\bar{X})\sum(y-\bar{Y})}{\sqrt{\sum(x-\bar{X})^2\sum(y-\bar{Y})^2}}$$

where

$$\sum(x-\bar{X}) = 106435.2,$$

$$\sum(y-\bar{Y}) = 106436.57,$$

$$\sum(x-\bar{X})^2 = 11328451799.1,$$

$$\sum(y-\bar{Y})^2 = 11328749819.6$$

$$r = \frac{106435.2 * 106436.57}{\sqrt{11328451799.1 * 11328749819.6}}$$

$$r = \frac{11328597615.27}{11328597615.22} = 1$$

The above correlation for hemorrhages is the positive correlation.

6.7.9 Statistical Techniques on Retinal Blood Vessels

Table 6.4 and Table 6.5

TABLE 6.4

Table for Retinal Blood Vessels Parameters

Sr. No.	Image Name	Area	Diameter	Length	Thickness	Mean Diameter	Tortuosity	Bifurcation Points	Venous Beading
01	image0092	17	13	9	2	19	2	117	Absent
02	image0093	22	15	11	2	20	10	127	Absent
03	image0094	15	12	8	2	19	12	88	Absent
04	image0095	25	16	13	2	19	8	233	Absent
05	image0096	15	12	8	2	19	2	112	Absent
06	image0097	17	13	9	2	19	2	138	Absent
07	image0098	24	16	12	2	20	1	136	Absent
08	image0099	12	11	6	2	20	1	90	Absent
09	image0100	17	13	9	2	19	5	112	Absent
10	image0101	15	12	8	2	19	14	93	Absent
11	image0102	47	22	24	2	20	2	275	Absent
12	image0103	25	16	13	2	19	2	131	Absent
13	image0104	22	15	11	2	20	2	163	Absent
14	image0105	16	13	8	2	20	3	130	Absent
15	image0106	21	15	11	2	19	3	176	Absent
16	image0107	10	10	5	2	20	2	56	Absent
17	image0108	16	13	8	2	20	2	97	Absent
18	image0109	21	15	11	2	19	2	119	Absent
19	image0110	24	16	12	2	20	2	155	Absent
20	image0111	23	15	12	2	19	6	186	Absent
21	image0112	21	15	11	2	19	3	117	Absent
22	image0113	14	12	7	2	20	10	93	Absent
23	image0114	16	13	8	2	20	2	141	Absent
24	image0115	28	17	14	2	20	3	142	Absent
25	image0116	12	11	6	2	20	6	92	Absent
26	image0117	23	15	12	2	19	4	136	Absent
27	image0118	26	16	13	2	20	4	142	Absent
28	image0119	14	12	7	2	20	9	128	Absent
29	image0120	19	14	10	2	19	2	161	Absent
30	image0121	12	11	6	2	20	2	104	Absent

TABLE 6.5
Table for Correlation of Diameter and Tortuosity of Retinal Blood Vessels

Sr. No.	Image Name	(x)	(y)	$(x-\bar{X})$	$(y-\bar{Y})$	xy
01	image0092	13	2	12.53	1.86	26
02	image0093	15	10	14.53	9.86	150
03	image0094	12	12	11.53	11.86	144
04	image0095	16	8	15.53	7.86	128
05	image0096	12	2	11.53	1.86	24
06	image0097	13	2	12.53	1.86	26
07	image0098	16	1	15.53	0.86	16
08	image0099	11	1	10.53	0.86	11
09	image0100	13	5	12.53	4.86	65
10	image0101	12	14	11.53	13.86	168
11	image0102	22	2	21.53	1.86	44
12	image0103	16	2	15.53	1.86	32
13	image0104	15	2	14.53	1.86	30
14	image0105	13	3	12.53	2.86	39
15	image0106	15	3	14.53	2.86	45
16	image0107	10	2	9.53	1.86	20
17	image0108	13	2	12.53	1.86	26
18	image0109	15	2	14.53	1.86	30
19	image0110	16	2	15.53	1.86	32
20	image0111	15	6	14.53	5.86	90
21	image0112	15	3	14.53	2.86	45
22	image0113	12	10	11.53	9.86	120
23	image0114	13	2	12.53	1.86	26
24	image0115	17	3	16.53	2.86	51
25	image0116	11	6	10.53	5.86	66
26	image0117	15	4	14.53	3.86	60
27	image0118	16	4	15.53	3.86	64
28	image0119	12	9	11.53	8.86	108
29	image0120	14	2	13.53	1.86	28
30	image0121	11	2	10.53	1.86	22

where

(x) = Diameter & (y) = Tortuosity

$$\text{Mean}(x) = \frac{13.96666667}{30} = 0.465555667$$

$$\text{Mean}(y) = \frac{4.266666667}{30} = 0.142222223$$

$$\text{Variance}(x) = \frac{\sum(x-\bar{X})}{N} = \frac{13.501111}{30} = 0.450037033$$

$$\text{Variance}(\text{y}) = \frac{\sum(y - \bar{Y})}{N} = \frac{4.124444443}{30} = 0.137481481$$

Standard Deviation (x): $\sqrt{Variance(x)} = \sqrt{0.450037033} = 0.670847996$

Standard Deviation (y): $\sqrt{Variance(y)} = \sqrt{0.137481481} = 0.370784953$

Correlation:

$$r = \frac{\sum(x - \bar{X})\sum(y - \bar{Y})}{\sqrt{\sum(x - \bar{X})^2 \sum(y - \bar{Y})^2}}$$

where

$$\sum(x - \bar{X}) = 13.50,$$

$$\sum(y - Y) = 4.12,$$

$$\sum(x - \bar{X})^2 = 182.28,$$

$$\sum(y - \bar{Y})^2 = 17.01$$

$$r = \frac{13.50 * 4.12}{\sqrt{182.28 * 17.01}}$$

$$r = \frac{52.62}{55.68} = 0.95$$

The tortuosity and diameter of retinal blood vessels are positively correlated.

6.7.10 Grading NPDR Lesions Using ANN

For grading of NPDR, ANN is applied. In machine learning and related fields, ANNs are computational models motivated by central nervous systems, in particular the brain, which is capable of solving complex machine learning as well as pattern recognition problems. ANNs are generally presented as systems of interconnected "neurons" which can compute values from inputs. In an ANN, simple artificial nodes, known as "neurons," "neurodes," "processing elements," or "units," are connected together to form a network which simulates a biological neural network.

For the detection of NPDR lesions, we designed graphical user interface using MATLAB 2013a.

In Figure 6.7, we have designed a GUI for the detection of NPDR lesions. Here, we have taken six axes: The first axes is used to read the original image and the second axes is used to display output of preprocessing operations of digital image processing techniques. After preprocessing, symlet wavelet is used for lesion detection. On axis 3, we extract MA

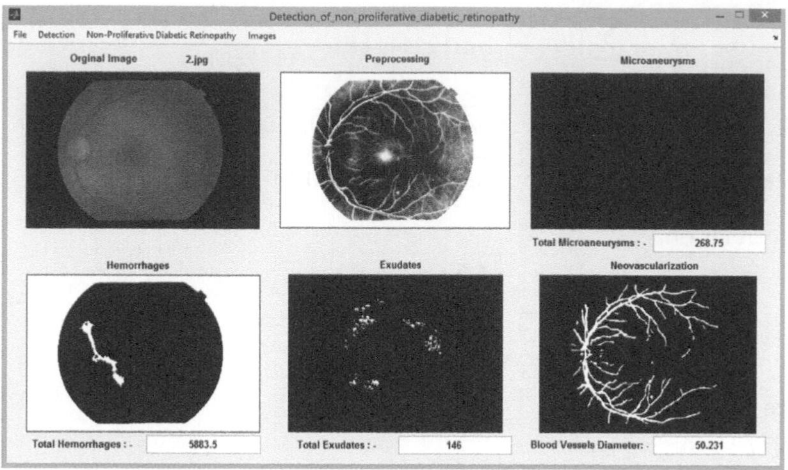

FIGURE 6.7
GUI for detection of NPDR lesions.

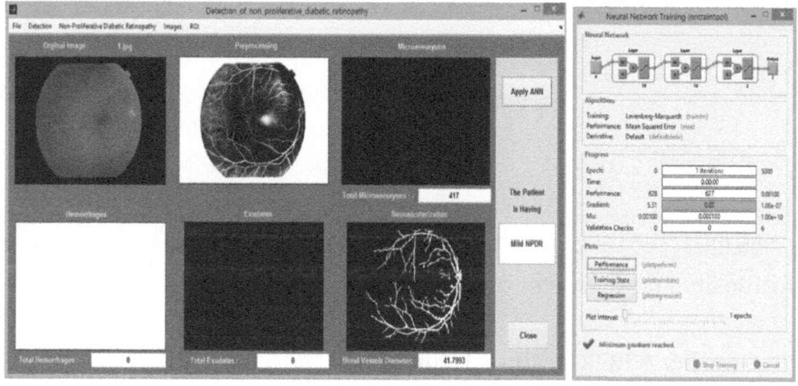

FIGURE 6.8
Mild NPDR lesions.

lesions, then on axis 4 we extract the hemorrhages. Subsequently, on axis 5, EXs are extracted, and on the last axis (axis 6) we extracted the retinal blood vessels. After the extraction of lesions, grading is very important, so for the grading of NPDR we use ANN.

Figures 6.8, 6.9, and 6.10 show the graphical user interface for NPDR lesions followed by the model of ANN for grading lesions in categories such as mild, moderate, and severe NPDR.

6.7.11 K-Means Clustering

K-means is a clustering algorithm. Clustering algorithms are unsupervised techniques for sub-dividing a larger dataset into smaller groups. The term

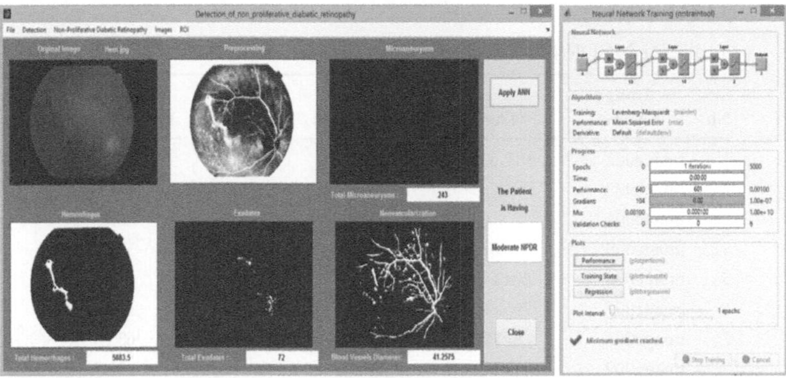

FIGURE 6.9
Moderate NPDR lesions.

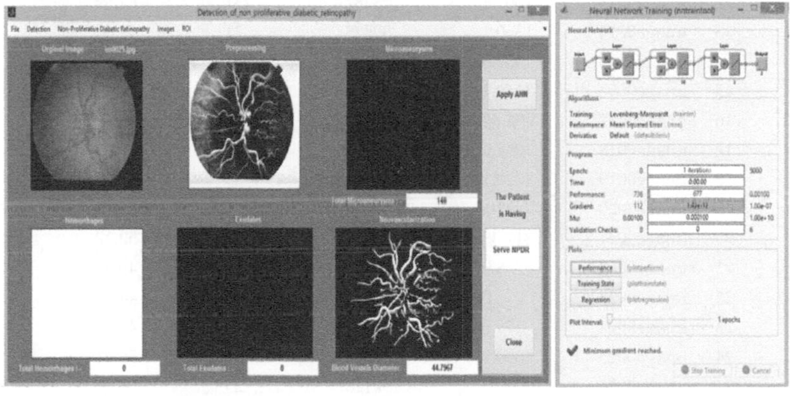

FIGURE 6.10
Serve NPDR lesions.

"unsupervised" indicates that the data do not originate from clearly defined groups that can be used to label them a priori. In machine learning, the problem of unsupervised learning is that of trying to find hidden structure in unlabeled data. Since the examples given to the learner are unlabeled, there is no error or reward signal to evaluate a potential solution. This distinguishes unsupervised learning from supervised learning and reinforcement learning. We apply k-means clustering on the following table.

The following figure shows the k-means clustering and k-means clustering after fitting.

> After removing a certain weight from the above data (Table 6.6), we exact clustered data in a grouped manner with normal in blue dots and abnormal in red dots. Figure 6.11 is clustered data but not purely classified, and after fitting the cluster we got pure classification as shown in Figure 6.12.

TABLE 6.6
Table for NPDR Lesions

Sr. No	Image Name	MAs	Hemorrhages	EXs	Diameter of Blood Vessels
1	image002	912.63	101400	17	45.32
2	image003	985.75	87659	26	46.74
3	image004	0	0	0	40.45
4	image005	905.38	116442	49	55.27
5	image006	2043.6	85989	152	51.41
6	image007	795.88	115240	70	51.65
7	image008	891.5	146930	29	63.36
8	image009	1138.8	92224	42	53.5
9	image0010	1347.1	78569	66	52.17
10	image0011	1688	77495	34	54.91
11	image0012	1474.9	76690	21	51.35
12	image0013	1005.6	78287	131	62.59
13	image0014	1360.1	89828	21	49.86
14	image0015	1437.5	78455	51	54.81
15	image0016	1149.1	85909	176	70.65
16	image0017	31.63	79424	74	49.78
17	image0018	1084.9	90696	38	47.45
18	image0019	1020.1	125990	3	51.03
19	image0020	734.63	81286	6	57.73
20	image0021	817.38	126310	85	50.27
21	image0022	1238.9	77106	18	49.61
22	image0023	0	0	0	44.57
23	image0024	276.5	152060	15	46.15
24	image0025	1680.9	78494	65	55.04
25	image0026	534	152120	25	51.33
26	image0027	0	0	0	43.94
27	image0028	1335.3	104120	36	49.31
28	image0029	785.88	144340	41	45.41
29	image0030	820.63	104880	26	58.33
30	image0031	902.38	133007	24	51
31	image0032	1183.8	90407	84	63.98
32	image0033	597.63	159450	35	51.22
33	image0034	331.63	234170	47	45.37
34	image0035	1398.3	78954	106	57.02
35	image0036	578.25	191393	31	46.51
36	image0037	1135.3	81948	68	52.58
37	image0038	847.13	111250	18	50.45
38	image0039	1003.4	111200	41	55.52
39	image0040	800.75	94149	96	58.09
40	image0041	984.5	95461	89	55.05
41	image0042	771.75	78084	48	51.16
42	image0043	739.5	132417	21	55.42
43	image0044	551.25	131140	48	50.98

(Continued)

TABLE 6.6 (CONTINUED)

Table for NPDR Lesions

Sr. No	Image Name	MAs	Hemorrhages	EXs	Diameter of Blood Vessels
44	image0045	690	92172	309	63.15
45	image0046	1114.1	88546	37	63.23
46	image0047	964.25	112680	66	55.86
47	image0048	1217.4	116630	65	55.58
48	image0049	824.13	79662	39	45.47
49	image0050	1087.1	103630	154	65
50	image0051	893.88	97124	8	50.29
51	image0052	1112.3	82670	37	51.72
52	image0053	1526.1	160620	10	47.84
53	image0054	0	0	0	44.53
54	image0055	687.88	189030	13	49.59
55	image0056	86.13	297683	5	51.96
56	image0057	809	126654	43	45.21
57	image0058	1610.8	98708	51	50.66
58	image0059	1006.1	150050	5	50.12
59	image0060	562.25	150070	31	47.63
60	image0061	1579.9	99959	41	50.48
61	image0062	641.13	168050	22	52.1
62	image0063	751.75	120070	61	49.03
63	image0064	1092.4	110070	99	55.19
64	image0065	677.75	131600	24	51.81
65	image0066	1107.5	145410	62	45.11
66	image0067	507.5	138320	20	52.94
67	image0068	1272.3	98868	44	50.16
68	image0069	956.38	80622	16	46.92
69	image0070	860.38	90982	45	45
70	image0071	968.25	113110	5	55.23
71	image0072	1335.3	104120	36	49.31
72	image0073	1064.5	98650	42	47.17
73	image0074	1501	82427	27	45.49
74	image0075	871.63	121290	33	56.13
75	image0076	741	116040	26	54.43
76	image0077	1740.9	80753	55	47.27
77	image0078	0	0	0	43.93
78	image0079	1075.8	87198	66	45.34
79	image0080	424	193540	7	48.79
80	image0081	1396.3	109350	35	48.66
81	image0082	0	0	0	43.25
82	image0083	1318.3	121640	26	53
83	image0084	463.38	183450	33	47.16
84	image0085	1112.5	75502	91	47.91
85	image0086	1015.4	75537	62	49.27
86	image0087	1178.3	95540	80	49.2

(Continued)

TABLE 6.6 (CONTINUED)

Table for NPDR Lesions

Sr. No	Image Name	MAs	Hemorrhages	EXs	Diameter of Blood Vessels
87	image0088	839	108640	57	53.23
88	image0089	1469.9	94552	63	52.17
89	image0090	1193.9	102800	50	55.07
90	image0091	1259.5	105220	58	49.81
91	image0092	1134.3	113610	51	46.38
92	image0093	910.5	124680	52	51.78
93	image0094	1382.8	95689	56	53.37
94	image0095	1206.3	119660	26	52.54
95	image0096	658.38	131565	37	48.24
96	image0097	334.5	211820	47	47.54
97	image0098	1110.6	116780	43	56.59
98	image0099	843.5	135890	71	48.06
99	image00100	1349.5	98019	56	56.47
100	image00101	1209.9	98823	47	61.68
101	image00102	156.88	209545	2	51.32
102	image00103	635	93746	24	49.38
103	image00104	1328.1	103850	68	50.28
104	image00105	984.13	100390	53	49.74
105	image00106	1653.4	78011	58	53.43
106	image00107	809.5	130530	51	43.93
107	image00108	1218.5	97040	78	59.67
108	image00109	1514.8	94163	67	62.79
109	image00110	794.63	95458	40	51.41
110	image00111	1301	81680	31	44.54
111	image00112	0	0	0	39.69
112	image00113	562	188770	14	46.89
113	image00114	0	0	0	39.19
114	image00115	637.88	194410	26	46.14
115	image00116	0	0	0	44.41
116	image00117	0	0	0	39.06
117	image00118	747.13	105190	57	45.53
118	image00119	1302.8	89258	75	49.07
119	image00120	1363.9	91106	44	47.93
120	image00121	981.5	124220	39	56.81
121	image00122	949.75	87920	90	53
122	image00123	0	0	0	44.65
123	image00124	0	0	0	44.83
124	image00125	1307.4	87369	67	49.17
125	image00126	1330.3	78071	19	46.09
126	image00127	224.13	125620	5	50.91
127	image00128	1072.1	91797	19	48.6
128	image00129	970.63	143370	43	49.85
129	image00130	1641.4	84380	121	53.48

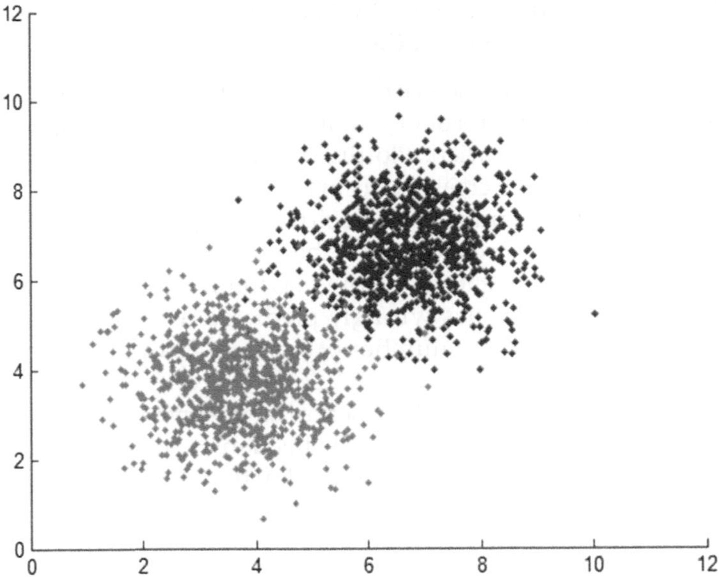

FIGURE 6.11
K-means clustering.

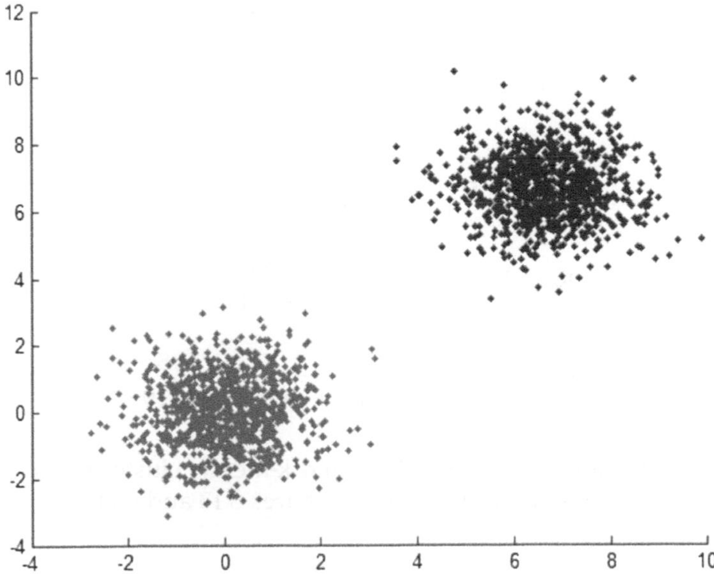

FIGURE 6.12
K-means clustering after fitting.

6.7.12 Performance Measurement by Receiver Operating Characteristic Curve

The receiver-operating characteristic (ROC), or ROC curve, is a graphical plot that illustrates the performance of a binary classifier system as its discrimination threshold is varied. Plotting the TP rate against the FP rate at various threshold settings creates the curve.

TP corresponds to the number of positive examples correctly predicted by the classifier. FN corresponds to the number of positive examples wrongly predicted as negative by the classifier. FP corresponds to the number of negative examples wrongly predicted as positive by the classifier. TN corresponds to the number of negative examples correctly predicted by the classifier.

The TP rate or sensitivity is the fraction of positive examples predicted correctly by the model. The FP rate is the fraction of negative examples predicted as a positive class. Precision is the fraction of records that actually turns out to be positive in the group the classifier has declared as a positive class. Recall is the fraction of positive examples correctly predicted by the classifier. F-measure is used to examine the tradeoff between recall and precision. Below are the calculations of receiver operating characteristic parameters.

$$\text{TP Rate} = \frac{TP}{(TP+FN)} = \frac{1180}{1170+11} = 0.99$$

$$\text{FP Rate} = \frac{FP}{(TN+FP)} = \frac{10}{0+10} = 1$$

$$\text{Precision} = \frac{TP}{(TP+FP)} = \frac{1180}{1180+10} = 0.99$$

$$\text{Recall} = \frac{TP}{(TP+FN)} = \frac{1180}{1180+11} = 0.99$$

$$\text{F - Measure} = \frac{2*TP}{(2*TP+FP+FN)} = \frac{2*1180}{(2*1180+10+11)} = 94$$

By using statistical techniques, ANN, and k-means clustering, it was concluded that the overall accuracy is 94% (Figures 6.13 and 6.14).

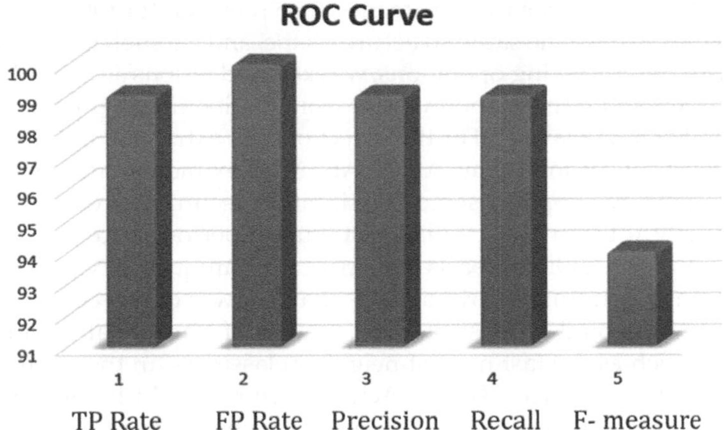

FIGURE 6.13
Performance analysis using ROC curve.

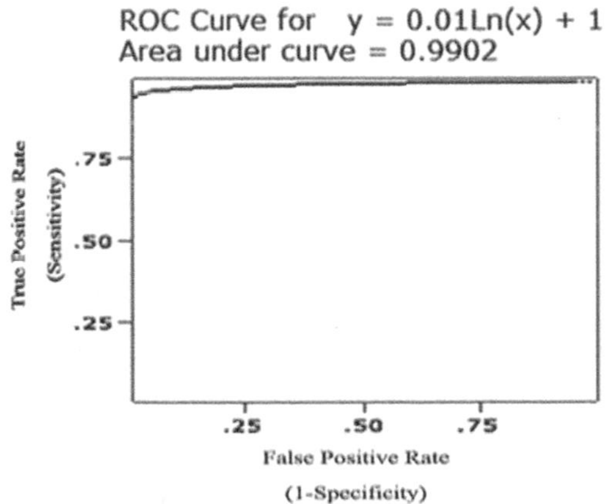

FIGURE 6.14
ROC curve.

6.8 Conclusion

NPDR is an early stage of diabetic retinopathy. In NPDR, tiny blood vessels within the retina leak blood or fluid and this leakage causes damage to the retina. NPDR is categorized in three stages, namely mild , moderate , and severe. For the detection of these NPDR lesions, firstly we downloaded online databases such as STARE, DRIVE, DiarectDB0, DiarectDB1, and SASWADE (the database collected during the research work). In total, we had 1,191 high-resolution fundus images. Before detecting NPDR lesions, we extracted the mask of the fundus images, then removed the OD because both EXs and the ODhave the same structure, shape, and geometry in some cases. After these preprocessing operations, we extracted lesions such as MAs, EXs, hemorrhages, and retinal blood vessels using digital image processing techniques and symlet wavelet. For the detection and extraction of these lesions we designed a GUI tool using MATLAB. After extraction and detection of NPDR lesions, we applied some classification techniques on NPDR data, using statistical techniques, ANN, ROC curve (performance analysis), and k-means clustering. On the basis of statistical techniques, ANN, ROC curve, and k-means clustering, an overall 94% accuracy was observed.

More studies could be possible with the use of traditional but improved classifiers, such as the fast nearest-neighbor classifiers (in the unsupervised framework) for clustering [1–47]. Active learning could be an interesting approach in case one needs to work on continuous data (data stream) [48]. High-level shape and texture features can be used to enhance the classification accuracy; a few of them can be taken from the previous works in chest imaging [49, 50, 51, 52, 53]. For complex images, fuzzy binarization is helpful, which is our immediate concern [54].

References

1. F. A. Jakobiec, *Ocular Anatomy*, Harper & Row Publisher, Inc, 1982.
2. Andrew A. Dahl and Thomas R. Gest, "Retina anatomy," *Medscape*. 2017. Available at https://emedicine.medscape.com/article/2019624-overview
3. Clara I. Sánchez, Roberto Hornero, María I. López, Mateo Aboy, Jesús Poza, Daniel Abásolo, "A novel automatic image processing algorithm for detection of hard exudates based on retinal image analysis," *Medical Engineering & Physics*, vol. 30, pp. 350–357, 2008.
4. Usman M. Akram and Shoab A. Khan, "Automated detection of dark and bright lesions in retinal images for early detection of diabetic retinopathy," *Journal of Medical Systems*, vol. 36, no. 5, pp. 3151–3162, 2011.
5. Arturo Aquino, Manuel Emilio Gegúndez, and Diego Marín, "Automated optic disc detection in retinal images of patients with diabetic retinopathy and risk of macular edema," *International Journal of Biological and Life Sciences*, vol. 8, p. 2, 2012.
6. Rangaraj M. Rangayyan, Xiaolu Zhu, Fábio J. Ayres, and Anna L. Ells, "Detection of the optic nerve head in fundus images of the retina with gabor filters and phase portrait analysis," *Journal of Digital Imaging*, vol. 23, no. 4, pp. 438–453, 2010.
7. Nathan Silberman, Kristy Ahlrich, Rob Fergus, and Lakshminarayanan Subramanian, "Case for automated detection of diabetic retinopathy," Copyright c 2010, Association for the Advancement of Artificial Intelligence (www.aaai.org). All rights reserved.

8. Sunrita Poddar, Bibhash Kumar Jha, and Chandan Chakraborty, "Quantitative clinical marker extraction from colour fundus images for non-proliferative diabetic retinopathy grading," in 2011 *International Conference on Image Information Processing (ICIIP 2011)*, 2011, 978-1-61284-861-7/11/$26.00 ©2011. IEEE.
9. R. Murugan and Reeba Korah, "An automatic screening method to detect optic disc in the retina," *International Journal of Advanced Information Technology (IJAIT)*, vol. 2, no. 4, 2012.
10. S. Kavitha and K. Duraiswamy, "Automatic detection of hard and soft exudates in fundus images using color histogram thresholding," *European Journal of Scientific Research*, vol. 48, no. 3, pp. 493–504, 2011, ISSN 1450-216X.
11. V. Vijayakumari and N. Suriyanarayanan, "Exudates detection methods in retinal images using image processing techniques," *International Journal of Scientific & Engineering Research*, vol. 1, no. 2, p. 1, 2010, ISSN 2229–5518.
12. Neera Singh and Ramesh Chandra Tripathi, "Automated early detection of diabetic retinopathy using image analysis techniques," *International Journal of Computer Applications*, vol. 8, no. 2, p. 18, 2010.
13. V. L. Rupa and P. S. Kulkarni, "Automatic diagnosis of diabetic retinopathy by hybrid multilayer feed forward neural network," *International Journal of Science, Engineering and Technology Research*, vol. 2, no. 9, 2013.
14. C. I. Sanchez, A. Mayo, M. Garcia, M. I. Lopez, and R. Hornero, "Automatic image processing algorithm to detect hard exudates based on mixture models," in Engineering in Medicine and Biology Society, EMBS '06, *28th Annual International Conference of the IEEE*, 2006, pp. 4453–4456.
15. N. Kingsbury, 4f8 image coding course. Lecture Notes, 2006.
16. Jack J. Kanski, *Clinical Ophthalmology: A Systematic Approach* (3rd ed.). Saunders, Philadephia, PA.
17. Rafael C. Gonzalez and Richard E. Woods, *Digital Image Processing*. Pearson Education, New Delhi, India, 2002.
18. Xiaolu Zhu, Rangaraj M. Rangayyan, and Anna L. Ells, *Digital Image Processing for Ophthalmology: Detection of the Optic Nerve Head*.
19. S. Jiméneza, P. Alemanya, I. Fondónb, A. Foncubiertab, B. Achab, and C. Serranob, "Automatic detection of vessels in color fundus images," *Sociedad Española de Oftalmología*, 2009.
20. Ahmed Wasif Reza, C. Eswaran, and Subhas Hati, "Diabetic retinopathy: a quadtree based blood vessel detection algorithm using RGB components in fundus images," *ACM Computing Surveys*, vol. 85, no. 3, pp. 103–109, 2010, 2007.
21. S. Vijayachitra, M. Menagadevi, and M. Ponni Bala, "Analysis of diabetic retinopathy images using blood vessel extraction," *International Journal of Advanced Engineering Research and Studies*, vol. I, number II. pp. 89–91, 2012, E-ISSN2249–8974.
22. Manjiri B. Patwari, Ramesh R. Manza, Yogesh M. Rajput, Manoj Saswade, and Neha K. Deshpande, "Calculation of retinal blood vessels tortuosity by using image processing techniques and statistical techniques," 2nd International Conference on System Modeling & Advancement in Research Trends (SMART) Department of Computer Applications, TMIMT, Teerthanker Mahaveer University, published in Academic Journal online (AJO), *International Journal of Trends in Computer Science*, vol. 2, no. 11, pp. 7462–8452, 2013.

23. Manjiri B. Patwari, Ramesh R. Manza, Yogesh M. Rajput, Manoj Saswade, and Neha K. Deshpande, "Detection and counting the microaneurysms using image processing techniques," *International Journal of Applied Information Systems*, vol. 6, no. 5, pp. 11–17, 2013. Published by Foundation of Computer Science, New York, USA, ISSN: 2249-0868, vol. 6, no. 5, October 2013.
24. Manjiri B. Patwari, Ramesh R. Manza, Yogesh M. Rajput, Manoj Saswade, and Neha K. Deshpande, "Automatic detection of retinal venous beading and tortuosity by using image processing techniques," *International Journal in Computer Application (IJCA)*, 2014, ISBN: 973-93-80880-06-7.
25. Manjiri B. Patwari, Ramesh R. Manza, Manoj Saswade, and Neha Deshpande, "A critical review of expert systems for detection and diagnosis of diabetic retinopathy," *CiiT International Journal of Fuzzy Systems*, 2012FS022012001, ISSN 0974-9721, 0974-9608. (IF 0.441).
26. Yogesh M. Rajput, Ramesh R. Manza, Manjiri B. Patwari, and Neha Deshpande, "Retinal blood vessels extraction using 2D median filter," in *Third National Conference on Advances in Computing* (NCAC-2013), School of Computer Sciences, North Maharashtra University, Jalgaon, India, March 5–6, 2013.
27. Yogesh M. Rajput, Ramesh R. Manza, Manjiri B. Patwari, and Neha Deshpande, "Retinal optic disc detection using speed up robust features," in National Conference *on* Computer *&* Management Science [*CMS-13*], Radhai Mahavidyalaya, Auarngabad, India, April 25–26, 2013.
28. Manjiri B. Patwari, Ramesh R. Manza, Yogesh M. Rajput, Manoj Saswade, and Neha K. Deshpande, "Review on detection and classification of diabetic retinopathy lesions using image processing techniques," *International Journal of Engineering Research & Technology (IJERT)*, vol. 2, no. 10, 2013, ISSN: 2278–0181.
29. Manjiri B. Patwari, Ramesh R. Manza, Yogesh M. Rajput, Neha K. Deshpande, and Manoj Saswade, "Extraction of the retinal blood vessels and detection of the bifurcation points," *International Journal in Computer Application (IJCA)*, 2013, ISBN: 973-93-80877-61-7.
30. Manjiri B. Patwari, Ramesh R. Manza, Yogesh M. Rajput, Manoj Saswade, and Neha K. Deshpande, "Persona identification algorithm based on retinal blood vessels bifurcation," in *2014 International Conference on Intelligent Computing Applications*, IEEE, 2014, 978-1-4799-3966-4/14.
31. Manjiri B. Patwari, Ramesh R. Manza, Yogesh M. Rajput, Deepali D. Rathod, Manoj Saswade, and Neha Deshpande, "Classification and calculation of retinal blood vessels parameters," in IEEE's *International Conferences for Convergence of Technology*, Pune, India.
32. Deepali D. Rathod, Ramesh R. Manza, Yogesh M. Rajput, Manjiri B. Patwari, Manoj Saswade, and Neha Deshpande, "Localization of optic disc using HRF database," in IEEE's *International Conferences for Convergence of Technology*, Pune, India.
33. M. Sezgin and B. Sankur, "Survey over image thresholding techniques and quantitative performance evaluation," *Journal of Electronic Imaging*, vol. 13, no. 1, pp. 146–165, 2004.
34. S. Chaudhuri, S. Chatterjee, N. Katz, M. Nelson, and M. Goldbaum, "Detection of blood vessels in retinal images using two-dimensional matched filters," *IEEE Transactions on Medical Imaging*, vol. 8, no. 3, pp. 263–269, 1989.

35. X. Jiang and D. Mojon, "Adaptive local thresholding by verification based multithreshold probing with application to vessel detection in retinal images," *IEEE Transactions on Pattern Analysis and Machine Intelligence*, vol. 25, no. 1, pp. 131–137, 2003.
36. J. Staal, M. D. Abramoff, M. Niemeijer, M. A. Viergever, and B. van Ginneken, "Ridge-based vessel segmentation in color images of the retina," *IEEE Transactions on Medical Imaging*, vol. 23, no. 4, pp. 501–509, 2004.
37. J. V. B. Soares, J. J. G. Leandro, R. M. Cesar, H. F. Jelinek, and M. J. Cree, "Retinal vessel segmentation using the 2-D gabor wavelet and supervised classification," *IEEE Transactions on Medical Imaging*, vol. 25, no. 9, pp. 1214–1222, 2006.
38. J. Nayak, P. Subbanna Bhat, U. Rajendra Acharya, C. M. Lim, and M. Kagathi, "Automated identification of diabetic retinopathy stages using digital fundus images," *Journal of Medical Systems*, vol. 32, pp. 107–115, 2008.
39. C. Sinthanayothin, V. Kongbunkiat, S. Phoojaruenchanachain, and A. Singlavanija, "Automated screening system for diabetic retinopathy," in *Proceedings of the 3rd International Symposium on Image and Signal Processing and Analysis*, 2003, pp. 915–920.
40. M. Niemeijer, M. D. Abramoff, and B. V. Ginneken, "Information fusion for diabetic retinopathy CAD in digital color fundus photographs," *IEEE Transactions on Medical Imaging*, vol. 28, no. 5, pp. 775–785, 2009.
41. T. Walter, J.-C. Klein, P. Massin, and A. Erginay, "A contribution of image processing to the diagnosis of diabetic retinopathy-detection of exudates in color fundus images of the human retina." *IEEE Transactions on Medical Imaging*, vol. 21, no. 10, pp. 1236–1243, 2002.
42. A. W. Reza, C. Eswaran, and K. Dimyati, "Diagnosis of diabetic retinopathy: automatic extraction of optic disc and exudates from retinal images using marker-controlled watershed transformation," *Journal of Medical Systems*, vol. 35, no. 6, pp. 1491–1501, 2010.
43. Manjiri Patwari, Ramesh Manza, Yogesh Rajput, Manoj Saswade, and Neha Deshpande, "Automated localization of optic disk, detection of microaneurysms and extraction of blood vessels to bypass angiography," *Advances in Intelligent Systems and Computing*, 2014, ISBN: 978–3-319-11933-5.
44. Karbhari Kale, Ramesh R. Manza, Ganesh R. Manza, Vikas T. Humbe, and Pravin L. Yannawar, *Understanding MATLAB*, Shroff Publisher & Distributer Pvt. Ltd., Navi Mumbai, 2013. ISBN: 9789350237199.
45. Ramesh Manza, Manjiri Patwari, and Yogesh Rajput, *Understanding GUI Using MATLAB*, Shroff Publisher & Distributer Pvt. Ltd., 2015, ISBN: 9789351109259.
46. Manjiri B. Patwari, Ramesh R. Manza, Yogesh M. Rajput, Manoj Saswade, and Neha K. Deshpande, "Personal identification algorithm based on retinal blood vessels bifurcation," in *2014 International Conference on Intelligent Computing Applications*, IEEE, 2014, 978-1-4799-3966-4/14.
47. S. Vajda and K. C. Santosh, "A fast k-nearest neighbor classifier using unsupervised clustering," in K. Santosh, M. Hangarge, V. Bevilacqua, and A. Negi (eds), *Recent Trends in Image Processing and Pattern Recognition*, 2017. RTIP2R 2016. Communications in Computer and Information Science, vol. 709. Springer, Singapore.
48. M. R. Bouguelia, S. Nowaczyk, , K. C. Santosh *et al.* "Agreeing to disagree: active learning with noisy labels without crowdsourcing," *International Journal of Machine Learning and Cybernetics*, vol. 9, p. 1307, 2018.

49. K. C. Santosh and Sameer K. Antani, "Automated chest X-ray screening: can lung region symmetry help detect pulmonary abnormalities?" *IEEE Transactions on Medical Imaging*, vol. 37, no. 5, pp. 1168–1177, 2018.
50. Fatema Tuz Zohora and K. C. Santosh, "Foreign circular element detection in chest X-rays for effective automated pulmonary abnormality screening," *International Journal of Computer Vision and Image Processing*, vol. 7, no. 2, pp. 36–49, 2017.
51. Alexandros Karargyris, Jenifer Siegelman, Dimitris Tzortzis, Stefan Jaeger, Sema Candemir, Zhiyun Xue, K. C. Santosh, Szilárd Vajda, Sameer K. Antani, Les R. Folio, and George R. Thoma, "Combination of texture and shape features to detect pulmonary abnormalities in digital chest X-rays," *International Journal of Computer Assisted Radiology and Surgery*, vol. 11, no. 1, pp. 99–106, 2016.
52. K. C. Santosh, Szilárd Vajda, Sameer K. Antani, and George R. Thoma, "Edge map analysis in chest X-rays for automatic pulmonary abnormality screening," *International Journal of Computer Assisted Radiology and Surgery*, vol. 11, no. 9, pp. 1637–1646, 2016.
53. K. C. Santosh, Sema Candemir, Stefan Jäger, Alexandros Karargyris, Sameer K. Antani, George R. Thoma, and Les R. Folio, "Automatically detecting rotation in chest radiographs using principal rib-orientation measure for quality control," *International Journal of Pattern Recognition and Artificial Intelligence*, vol. 29, no. 2, 2015.
54. K. C. Santosh, Laurent Wendling, Sameer K. Antani, and George R. Thoma, "Overlaid arrow detection for labeling regions of interest in biomedical images," *IEEE Intelligent Systems*, vol. 31, no. 3, pp. 66–75, 2016.

7

Segmentation and Analysis of CT Images for Bone Fracture Detection and Labeling

Darshan D. Ruikar, K.C. Santosh, and Ravindra S. Hegadi

CONTENTS

7.1 Introduction

Developing computerized solutions for several orthopedic healthcare practices is one of the emerging fields in computer science. Development of expert systems such as CAD for fracture reduction surgery, automated preoperative surgical planners, outcome prediction systems, intra-operative assistant systems, and virtual reality (VR)-based simulators for surgical skill training are drastic needs for the orthopedic society. An expert system assists the surgeon in deciding the optimal recovery plan/process. Moreover, it helps to predict the outcome of surgery before treating the patient. A computer-assisted VR-based simulator provides a controlled, safe, and secure training environment to novice trainees to achieve expertise in their craft. Moreover, trainees can do repeated practice at no cost [1].

After patient-specific data collection in the form of medical images, the precise extraction of (healthy or fractured) bones from those images and assigning appropriate labels to them by considering bone anatomy are the mandatory steps in almost all the aforementioned applications. Accurate segmentation helps achieving an in-depth understanding of the severity of the injury, and also helps with the anatomically accurate visualization of the complex geometrical structure of the bone. In addition to this, accurate segmentation will result in several fracture features, such as the type of fracture, number of bones, number of fractured pieces per bone, and amount of dislocation of fracture piece from its original location. These features help surgeons to decide an optimal recovery plan [2].

However, devising a precise segmentation method is a challenging task. The low resolution of CT images, the presence of unwanted artifacts, intensity variation, and fuzzy fracture lines are some of the inherent challenges in CT imaging. In addition to this, complex bone anatomy, the severity of fractures, variation in size and shapes of fractured pieces, dislocation of fractured pieces, and misconnection of fractured pieces due to dislocations are additional challenges that need to be considered. In some complicated cases, past experience is not applicable and expert guidance is required to segment fractured bones accurately [2]. To overcome these challenges, and moreover to minimize user intervention the field of orthopedics, technological advancements need to be adopted.

In the proposed work, we have developed a CAD system for bone fracture detection and analysis. The proposed system works in stages: unwanted artifacts removal, bone region extraction, and unique label assignment. In the first step, the acquiesced CT images are preprocessed to remove unwanted artifacts and to enhance bone regions. A histogram modeling and point processing-based image enhancement technique are devised to erase the flesh surrounded by bone tissue, and to enhance bone tissue regions. A 2D region growing-based segmentation method is adopted to extract bone tissue regions from the preprocessed image. The seed points are selected automatically. In the last stage, the hierarchical labeling scheme is used to assign the unique labels to each fractured piece by considering patient-specific bone anatomy. The labeling scheme treats a number of individual bones and their fractured pieces individually while assigning labels. In addition to this, the proposed system provides following features such as number of bones in inspected CT stack and number of fractured pieces per bone. These features are beneficial to analyze and understand the severity of the injury. In addition to this, these features are helpful to devise an optimal recovery plan.

The organization of this chapter is as follows: Section 7.2 provides brief information about related clinical aspects such as anatomy of long bone, CT imaging,andvarious types of fractures. Detailed reviews of existing CT image segmentation algorithms and computer-based bone fracture analysis systemsare discussed in Section 7.3. Along with this, the merits and demerits of each technique are presented in the same section. Section 7.4 describes the design and development of the proposed CAD system by providing detailed information about preprocessing, segmentation, and label assignment techniques, which are employed for unwanted artifacts removal, bone fragment extraction, and uniquelabel assignment respectively. The experimental results of real patient-specific images and comparisons with the clinical ground truth are presented in Section 7.5. This section also provides the comparative result of several competing methods. Section 7.6 gives conclusions and directions for future work.

7.2 Clinical Aspects

7.2.1 Anatomy of Long Bone

According to human anatomy, long bones such as femur* or humerus† can be divided into two regions: diaphysisand epiphysis, as shown in Figure 7.1. The middle portion of a bone is known as diaphysis or shaft, whereas the end portion is known as epiphysis. Furthermore, the upper end is called proximal epiphysis, and the lower end is known as distal epiphysis. The bone is majorly made up of two types of tissues: cortical tissue and cancellous tissue. Cortical tissues are present in the outer part of the bone. They are dense and provide strength to the bone, whereas cancellous (trabecular) tissues are found in the inner part of the bone. They have a spongy structure and are helpful in the injury recovery process [3].

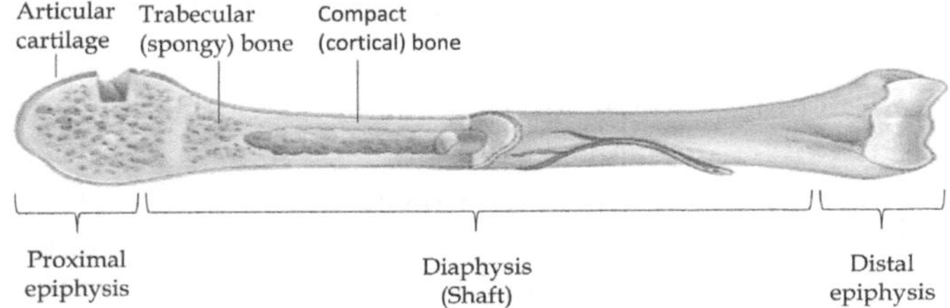

FIGURE 7.1
Anatomy of long bone.

* The femur bone is present in the lap,and runs from hip to knee joint.

† The humerus bone is present in the arm,and runs from shoulder to elbow.

7.2.2 CT Imaging

Medical imaging modalities such as X-ray and CT images are majorly used for bone trauma diagnosis and prognosis [4]. These images provide better visualization and precise understanding of bone injury. Nowadays, CT images are widely used as a medical diagnostic imaging modality to analyze and prognose bone injury. Though X-ray images are a better alternative concerning cost and radiation doses, CT images are used most often [5]. There are two primary reasons behind this: (1) the X-ray image formation mechanism is based on the mean absorption of various tissues through which X-rays are passed. Hence the quantitative measurement of individual tissue density and precise discrimination of various soft tissues is not possible. (2) X-ray images are 2D images, so accurate visualization of bone anatomy, perfect interpretation of Sevier bone injuries, and 3D reconstruction are practically difficult. Moreover, the depth information is also lost,and different organs may superimpose on each other andmislead the viewer [6].

To overcome the above-discussed limitations of X-ray imaging, a novel medical imaging technique named computerized axial transverse (CAT) or computerized tomography (CT) scanning was developed by Godfrey Hounsfield in 1972 [6]. CT images are formed by making use of computer and X-ray measurements taken from several angles to produce cross-sectional (tomographic) images. They provide clear insights into a scanned object without cutting. The formation of a CT image, i.e. slice, depends on the attenuation coefficients of the scanning material. The X-ray generators generating the X-ray beam are rotated around the human body and the X-ray detectors are mounted exactly in the opposite position, which computes the attenuation coefficient based on the density of scanning material. The density of a material is measured in terms of Hounsfield units, which are a linear transformation of the tissue attenuation coefficients. The Hounsfield scale varies according to the density values of the material. For air, water, and dense bone it is –1,000, 0, and 1,000 respectively [7].

A CT stack is a comprehensive collection number of axial CT slices. Hence, it accurately preserves the depth information, which is helpful in the 3D reconstruction process. Moreover, the formation of the CT image is based on the measurement of X-ray beam attenuation of individual tissue, so quantitative measurement, as well as discrimination amongst them,are possible [7].

In the formed CT image, each pixel is the gray intensityvalue proportional to the density of the tissue. Over the CT stack, both cortical and cancellous tissues show high variation in intensity values. That is, intensity values of the same tissue may not be the same in all the slices [2]. Figure 7.2 shows the variation in intensity values of cortical tissues in the same patient's CT stack. Near diaphysis, they are brighter and denseas shown in Figure 7.2(a), whereas as we move nearer to the joints they become thinner and appear very fuzzy, as shown in Figure 7.2(b). In some slices, they may disappear.

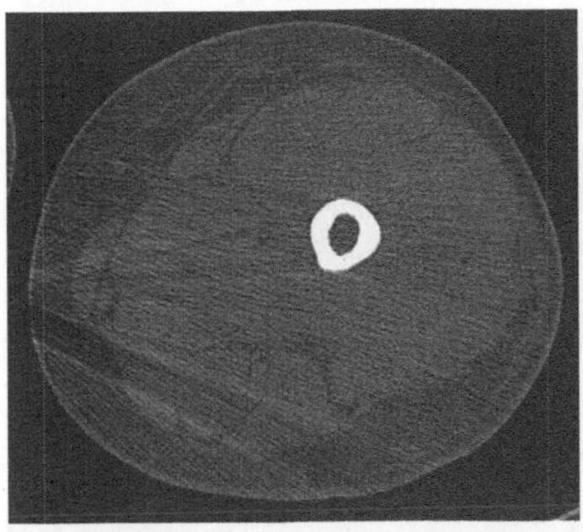
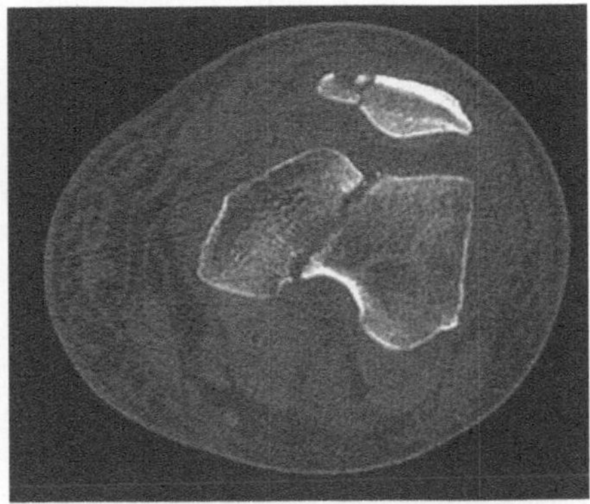

FIGURE 7.2
CT images of the same patient (a) diaphysis, (b) epiphysis.

In eight-bitgray-scale, image intensity values of cortical tissues range from 215 to 255, i.e. they are brighter, hence easily identifiable, whereas cancellous tissues range from 110 to 210. More generally, intensity values of cancellous tissues are lowerthan the CT range; moreover, these values are the same as the values of soft tissues.

7.2.3 Types of Fractures

A fracture is a broken bone. Fracture occurs when an outside force is applied. If the force is too high, the bones will break [8]. The severity of a fracture usually depends on the force that caused the break. In mostorthopedic traumatic cases or accidents, the joint area is more likely to fracture than the shaft, i.e. epiphysis is more prone to fracture than diaphysis. The epiphyses of two or more bones come together at the joint area; the layer of cortical tissues (which provides strength) is fragile. That is the main reason behind the fact that the entire housing of the joint area is more vulnerable to fracture. In case of fracture at the joint area, the fracture lines are not crisp and appear very fuzzy. Contrary to this, a shaft, i.e. the middle part of the bone, is less fracture prone. Breakage happens only when massive outside force is applied directly on it. In the case of fracture, fracture lines are crisp,and the identification of individual fracture pieces is easier because there is a higher presence of cortical tissues at the shaft. Sample CT images containing fractures at the joint and shaft areasareshown in Figure 7.3(a) and (b) respectively. Most often greenstick, transverse, oblique, spiral, or segmental fractures may appear at the shaft. Avulsed or comminuted fractures may appear at the nearby joint area.

In a greenstick fracture, one side of the bone is broken, and the other only bent. The fracture line is perpendicular to the shaft in a transverse fracture, whereas the fracture line makes an angle with the bone in an oblique fracture. A spiral fracture occurs when the body is in motion,and a rotating force is applied along the axis of a bone. An avulsed fracture occurs when a tendon

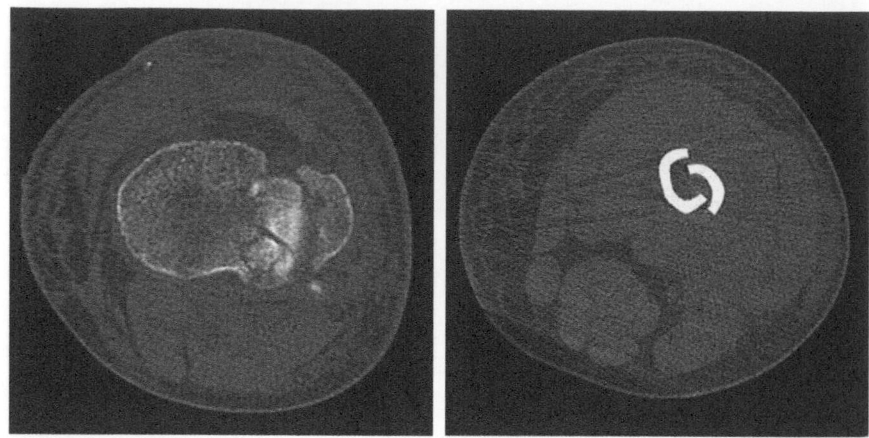

FIGURE 7.3
CT images of fracture near the joint area (a) and in the shaft area (b).

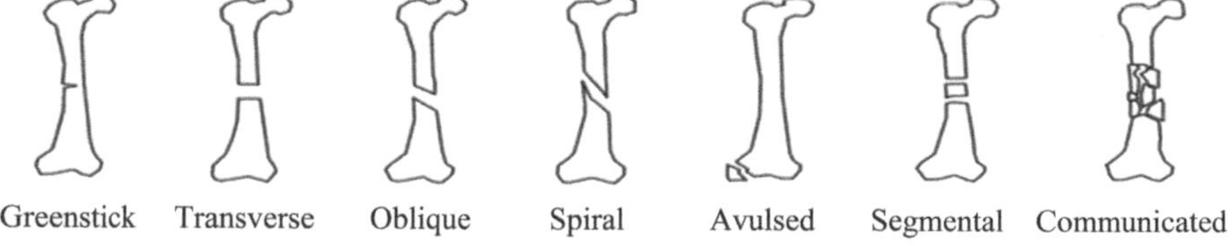

FIGURE 7.4
Types of fractures.

or ligament pulls off a piece of the bone. A segmental fracture contains at least two fracture lines with several separate pieces. If the bone is broken into multiple pieces, possibly with dislocation, then the fracture is called comminuted [3]. All of the fracture types mentioned are shown in Figure 7.4.

The accurate identification of the fracture line and individual pieces is a challenging task, especially in the presence of comminuted fractures at joints. Because of a thin layer of cortical tissues, the fracture line is not crisp. In addition to this, due to dislocation of fracture pieces, they may get wrongly connected to other pieces, which in turn leads to over- or under-segmentation [8].

7.3 Literature Survey

Segmentation is a process thatdivides the image into several parts depending on the need of the application [9]. It is the first and most crucial step in many image processing-based applications. The field of medical image analysis is also no exception to this. In orthopedic healthcare applications such as CAD system development, visualization, surgical planning, and simulation, segmentation plays an important role [5]. The eventual success of any application majorly depends on the result of segmentation. Hence, many successful research attempts are made by researchers to develop efficient segmentation techniques to extract the desired portion from medical images. Boundary-based, region-based, and atlas-guided thresholding are a few of the approaches used to segment healthy as well as fractured bones. Detailed information about the above-mentionedsegmentation techniques are discussed in the literature [5, 9].

The segmentation of healthy bone from CT images is a laborious task. The low resolution of CT images, the anatomy of the bone under supervision, and the intensity variation of the same bone tissue over the slices are some of the inherent challenges [10]. Hence, using a standard solution which will apply to every case is impractical. In addition to these innate difficulties, the complexity ofthe fracture, the number of pieces, amount of dislocation, and wrong connection of dislocated pieces are some of the additional challenges we need to consider to devise a solution for segmenting and labeling fracture bones [2]. Furthermore, past experience is not always applicable because it is very uncommon to find the same fracture case again.

Tomazevic et al. [11] proposed a thresholding-based interactive segmentation toolkit to extract a bone region from CT images. The kit is made up of several tools such as the separation tool, the merge tool, and the hole filling tool to segment individual fractured pieces. The tools can be operated in two ways: manual and semiautomatic. A global thresholding-based segmentation method is proposed in Tassani et al. [12] to detect several fractured regions in the cancellous tissue region. Due to the intensity inhomogeneities problem, it is difficult to find out a single threshold value that is applicable for the slices. Binarization of CT images of the humerus bone is done by using a reasonable threshold-based segmentation technique. Several morphological operations and Gabor wavelet transform are used to extract the cortical and trabecular tissues from an image [13]. Other than the thresholding-based segmentation method, maximum research attempts are made by adapting 2D [2,15] or 3D [16, 17] region growing-based segmentation methods to healthy segments as well as fracture-prone bones. This method requires a little user intervention in terms of seed point and threshold value selection.

In Shadid and Willis [2], a 2D region growing-based segmentation approach is used to segment and label fractured pieces from CT images. Initially, the curvature flow filter is used to enhance the bone boundaries. Then the user places several seed points (one per fractured piece) to extract the desired portion. Also, this proposed method expects extra seeds to separate erroneously connected fractured pieces. A multi-region growing-based segmentation method was proposed by Lee and et al. [18] to segment fractured bones from CT images. The seed points are identified automatically by scanned input image several times. Suppose that in some complicated cases, the automated approach fails: then the user can use region splitting and merging tools to separate and merge incorrect segmented regions respectively. To minimize user interaction and to improve segmentation accuracy a sheetness-based enhancement technique is used in several studies [14, 15]. The proposed enhancement technique efficiently identifies the cortical and cancellous bone tissues separately. Later, a 3D region growing-based segmentation method is adapted to extract fractured bone pieces from an image. In addition to this, an interactive graph cut method is used [14] to separate misconnected pieces, and a 3D connected labeling algorithm is applied to label each fractured zone. Along with thresholding and region growing methods, researchers also adapt graph cut [14], deformable models [19, 20], probabilistic watershed transforms [21, 22], registration [23], and fuzzy binarization [24]segmentation techniques to segment and label fractured bones.

In short, several segmentation methods are adapted by researchers to segment healthy as well as fractured bone. However, many research attempts are based on variations in thresholding and region growing techniques. Most of the previously developed methods are semiautomatic; they expect some user intervention to achieve the desired result. User interaction is in the form of threshold value or seed point selection. Several medical image analysis and visualization tools such as DICOM viewer, 3D slicer, InVesalius, and Dolphinare also used the interactive thresholding-based segmentation method to extract a bone part from animage [25, 26]. In addition to this, they require more user interaction to separate wrongly connected fractured components.

In the literature, most of the research attempts are made in an integrated environment where hospitals and engineering departments are under the same roof. No database is publically available for bone fracture detection and analysis purpose. Researches have used conventional image preprocessing methods; less importance is given to developing useful CT image preprocessing techniques which will remove unwanted artifacts and enhance bone regions by analyzing the image contents. In addition to this, existing labeling logic is also not robust. It has a few limitations: for instance, it does not consider bone anatomy while assigning labels to extracted bone regions. If a CT slice has several individual healthy bones, it may consider them as fractured pieces of one bone. Moreover, fractured pieces of different bones are treated differently.

In the presented work we have developed an effective preprocessing and labeling mechanism. The proposed artifact removal technique precisely analyses the image contents to remove the unwanted artifacts, enhances the bone tissue regions, and is based on histogram modeling and point processing-based image enhancement technique. The developed hierarchical labeling technique is also innovative and assigns unique labels to eachbone and their respective fractured pieces by considering bone anatomy.

7.4 Proposed Methodology

7.4.1 Data Acquisition

Patient-specific data collection is the initial step in CAD and surgery simulation system development. The hospitals and radiology centers all over the globe are enriched with biomedical data. Due to ethical laws such as HIPPA* and patient-specific data sharing protocols such as IRB†, they hesitate to share those data. If data were made available to researchers in other fields, this would be a beneficial scenario to enhance research and product development in the healthcare field [27, 28]. In the proposed work, 28 patient-specific CT stacks are collected from several hospitals and radiology centers in India. In order to test the robustness of the proposed system, data are deliberately collected from different sources that have CT scan machines with different configurations. Each CT stack is a comprehensive collection of 100 to 500 CT slices. The number of slices per CT stack mostly depends upon the thickness of the slice, the area under supervision, and the severity of the injury.

In total, 8,000 patient-specific CT slices are collected in the DICOM‡ format. DICOM files contain two sections: header and data. The header section contains patient-specific information such as patient number, name, and age. The data section contains a scanned portion. Before further processing, patient-specific information is a must. MATLAB script was developed to erase patient-specific information from the slice. The script uses the predefined functions in the image processing toolbox such as dicomanon, dicomread, and dicomwrite.

7.4.2 Data Annotation

To provide an expected result, i.e. clinical ground truth, the CT images are annotated by expert orthopedic surgeons. Experts are very busy in their routine tasks, and they do not have enough time to annotate every slice in a vast database. In addition to this, we observed that the same fracture scenario keeps on repeating at least for the slices with little variation in size or location of fractured pieces. Hence, every 10th slice in each patient-specific CT stack is annotated by an expert orthopedic surgeon. Figure 7.5 shows the annotated CT images in the same patient-specific CT stack. The ith and i+10th slices in the shaft region are shown in Figure 7.5 (a) and (b). This shows a small change in the location of the fractured piece, whereas the ith

* HIPPA: Health Insurance Portability and Accountability Act.
† IRB: Institutional Review Board.
‡ DICOM: Digital imaging and communications in medicine.

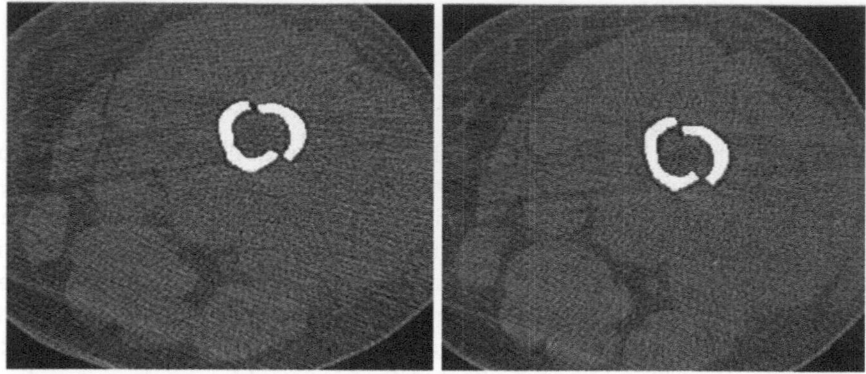

FIGURE 7.5
CT images with fracture at shaft (a) ith and (b) I +10th slice.

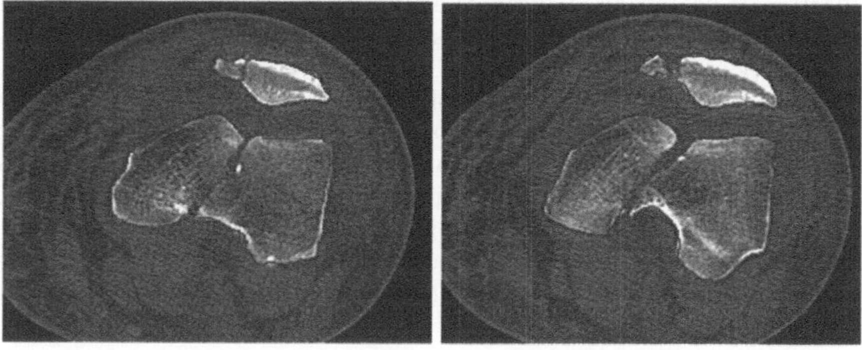

FIGURE 7.6
CT images with a fracture near the joint area (a) ith and (b) I +10th slice.

and i + 10th slices near the joint region are shown in Figure 7.6(a) and (b) respectively. This shows small growth in the size of the fractured piece.

The fracture is the discontinuities present at the outer cortex of the bone. During the annotation process, the boundaries of the bone regions are highlighted by experts. Small fractured pieces (with boundary pixels lower than 50) are discarded from the annotation process, because according to experts' opinion these pieces are removed while treating the patient.

7.4.3 Unwanted Artifacts Removal

In addition to the bone region, the CT image may contain some unwanted artifacts such as CT bed, cables, and flesh. To focus on the development of the CAD system for fracture detection and analysis, and to concentrate on the desired portion (fractured bone regions), the unwanted artifacts have to be removed first [29]. Commonly, the scanned portion is present at the center of the image; hence, the inputted image is cropped by 100 pixels from all directions (left, right, top, and bottom). This cropping operation results in the removal of the CT bed and cables. Now, to remove flesh that is surrounded by bone region, the histogram modeling and point processing-based image enhancement technique has been devised.

7.4.3.1 Histogram Stretching

The histogram provides a global description and enormous information about an image [30]. Hence, a histogram is plotted (Figure 7.7 (b)) for a sample CT image (Figure 7.7 (a)) to analyze the image contents. Histogram analysis results: the CT image is a low-contrast image as maximum pixels fall under small range. That is, a large number of pixels have intensity value ranging from 70 to 120. This happens because the flesh occupies most of the image area. To remove unwanted flesh and to enhance the bone tissue histogram, stretching is necessary.

Histogram stretching is a point processing-based image enhancement method used to increase the dynamic range of the image [31]. This method is responsible for spreading the histogram to cover the entire dynamic range without affecting the original shape of the histogram. The transformation function in Equation 7.1 is used to increase the dynamic range of the image by stretching the intensity range of the inputtedimage ($R_{min,}$ R_{max}) to cover the entire dynamic range ($S_{min,}$ S_{max}).

$$S = T(R) = (S_{max} - S_{min}) / (R_{max} - R_{min}) * (R - R_{min}) + S_{min}, \qquad (7.1)$$

where R is the input and S is a resultant image. The function T(R) is based on the equation of the straight line having a slope $(S_{max} - S_{min})/(R_{max} - R_{min})$. It

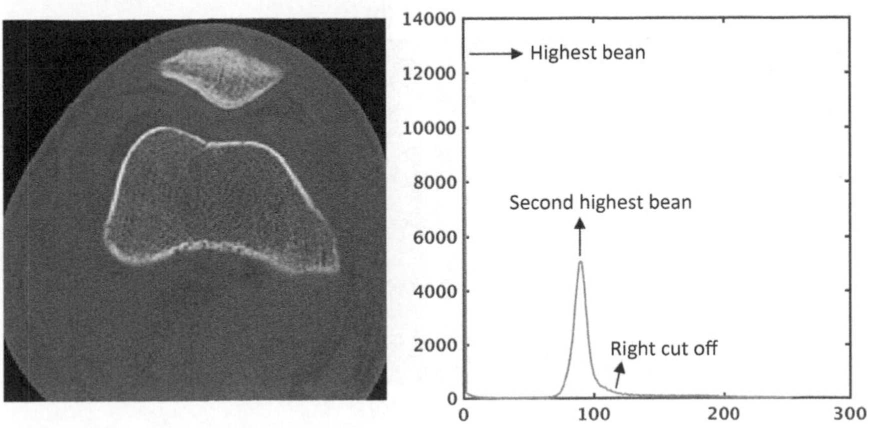

FIGURE 7.7
(a) CT image and (b) histogram.

accepts four parameters: S_{min}, S_{max}, R_{min}, and R_{max}. In an 8-bit image, the values of S_{min} and S_{max} are 0 and 255 respectively to occupy the entire dynamic range. To obtain image-specific intensity range (i.e. R_{min} and R_{max}), histogram plots are observed. The intensity value of the second highest bean is assigned to R_{min} (the first highest bean represents a number of black pixels) and the right cut-off second highest bean is assigned to R_{max}. After getting the values for all four parameters, the transformation function in Equation 7.1 is applied to stretch the narrow intensity range to cover the entire dynamic range. As a result of this, the flesh and the bone tissue that surrounds it are removed as the image is stretched to darker intensity values (i.e. intensity values nearer 0). The bone tissues are enhanced precisely as they are stretched to brighter intensity values (i.e. intensity values nearer 255). Figure 7.8 (a) and (c) shows the original CT image, whereas (b) and (d) show an enhanced image at the shaft and joint regions respectively. The resultant images (Figure 7.8 (b) and (d)) shows that the unwanted flesh is removed nicely after applying the proposed image enhancement technique.

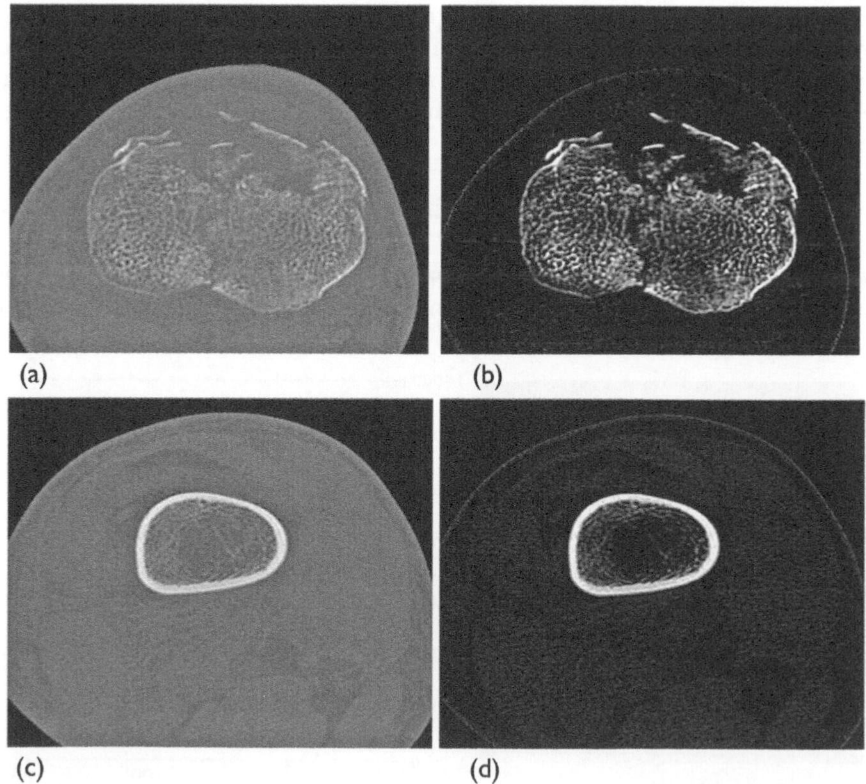

FIGURE 7.8
(a) and (b) original CT image; (c) and (d) enhanced images with unwanted artifacts removal.

7.4.4 Bone Region Extraction and Labeling

As the eventual results of the image processing–based application majorly depend on the results of the segmentation method [32,33], on the same note, the results of the segmentation method areprimarily based on the results of the image enhancement technique. If the enhancement technique successfully removes the unwanted part and precisely enhances the desired portion of the image, then a simple segmentation technique is enough to extract a required portion from the image. That is a promising (image content analysis–based) image enhancement method that leads to the adaption of a simple segmentation method. The above-discussed histogram stretching-based image enhancement technique effectively removes the unwanted flesh which is surrounded by bone; moreover, it enhances the bone tissue regions nicely. Hence, we adopted a simple 2D region growing-based segmentation method [34] to extract the required regions from the image. The 2D seeded region growing method expected two parameters from the user: seed points and threshold value. In the proposed work, both these parameters are identified automatically by analyzing the image contents. The following two sub-sections provide adetailed explanation on automatic seed points and threshold value selection.

7.4.4.1 Seed Point Selection and Spreading

To select a seed point, the first image in the CT stack is scanned from top to bottom to identify a first point whose intensity value is higher than 220 (initial value of cortical tissues). Algorithm 1 describes the seed point selection process. Then that point is considered as a seedpoint and starts getting spread by the 2D region growing algorithm. If the size of the current spread region is lowerthan 50, then the region and respective seed point are discarded (as discussed in Section 7.4.2, smaller bone fragments are not considered for further process and are removed from the fracture-prone area). If the region size is higher than 50, then that seed point is stored in a global list for future reference, and the currently spread region is subtracted from the CT image. The same process is executed repeatedly until no more regions are found. If there are more than one seed point in the global seed list during the seed selection process for the firstslice,then the number of individual bones exists in the current CT stack.

Algorithm 1: SeedPointSelection(I)

```
[x,y] = size(I);
for i = 1 to x do
  for j = 1 to y do
    if (I(i,j) ≥ 220) then
      return (i,j) as a seed point;
    end if
  end for
end for
```

After identifying all seed points for the firstslice, the same set of seed points is propagated for the rest of the slices. All the seed points are spread one after another, and the size of the current spread region is checked with the region that was spread by the same seed point in the previous slice. If the difference in both the region sizes is more than 20, then it may lead to one amongst two scenarios. A) The difference is positive (i.e. region size of the current slice is lesser then region size of the previous slice) then the bone fracture is introduced, and the bone is broken into more than one piece. B) The difference in size is negative (i.e. the current region size is larger than the previous one) then another fractured piece may get wrongly connected to the current bone due to dislocation or proximity of CT images. In the first case, to identify the new seed point for the fractured bone, the same seed point selection process is executed. For the second case, that is to separate the wrongly connected fractured pieces, morphological opening operation is performed. If, after spreading all the existing seed points and subtracting the respective spread regions, the inputted image is still not empty, then there exists new bone in that slice. The same seed point selection and the spreading process are repeated for that region, and the new seed point is placed in the global list.

7.4.4.2 Threshold Value Definition

The threshold value is another parameter of the seeded region growing algorithm. The same bone tissue shows variations in intensity values. This is possible for both cortical and cancellous tissues. The cortical tissues are brighter and thicker at the shaft regions, whereas they appear fuzzy and thinner near the joint region. Hence, it is difficult to set a single global threshold value that is applicable for the entire CT stack.

To select a threshold value, a 5×5 window is created by considering the seed pixel inthe top-left position in that window, and the average of that window is considered a threshold value to extract that region by seeded region growing algorithm. The reason behind window formation is that the optimal threshold value must be set by considering the adjacent elements.

7.4.4.3 Unique Label Assignment

While the seed selection and spreading process are being performed, unique labels are assigned to every individual bone and each fractured piece of the respective bones by considering bone anatomy. When scanning the long bones, generally a radiologist starts scanning from the unfractured region, then enters into the fracture-prone area, and continues scanning till he or she reaches a safe state (i.e. till unfractured bone region appears). Hence, it is very uncommon to get fractured pieces in the first slice itself. The proposed labeling technique assigns labels 10, 20, and so on to each extracted bone region from the first slice. The first digit in the label indicates a count of bones in that slice, and the second digit (i.e. 0) indicates healthy bone where no fracture has been introduced yet. Same labels and seed points from first slice are continued for upcoming slices. Along with seed point and label propagation process, the difference in region sizes are also calculated for subsequent slices. This seed point and label propagation process is continues till the difference in region sizes is more than 50. As discussed in Section 7.4.4.1, if the difference is positive then the fracture is introduced for the respective region. Then the proposed labeling technique changes the label of the currently extracted region to 11 to indicate that it is the first fracture piece. After that, it assigns labels 12, 13, and so on for newly extracted regions. Just for the sake of confirmation, any random point in the newly extracted region is considered as a test point and a search process is carried out to check whether that test point is present in a region which is extracted by seed point having label 10 in the previous slice. It is quite evident that the test point will be found in that region, as all those are fracture pieces of same bone. The same process is repeated for each seed in the global seed list and continues till it reaches the last slice in the CT stack.

The proposed labeling method is innovative and assigns unique labels to each bone region by considering patient-specific bone anatomy. That is, the method treats several bones and their fractured pieces differently-from assigning labels. Figure 7.9 (a) and (b) shows the ith and i + 1th in same patient-specific CT stack. In i + 1th, a fracture is introduced and the bone is divided into two pieces. The labeling logic will assign labels 11 and 12 to those pieces. The bone region extraction and labeling procedure is explained in Algorithm 2.

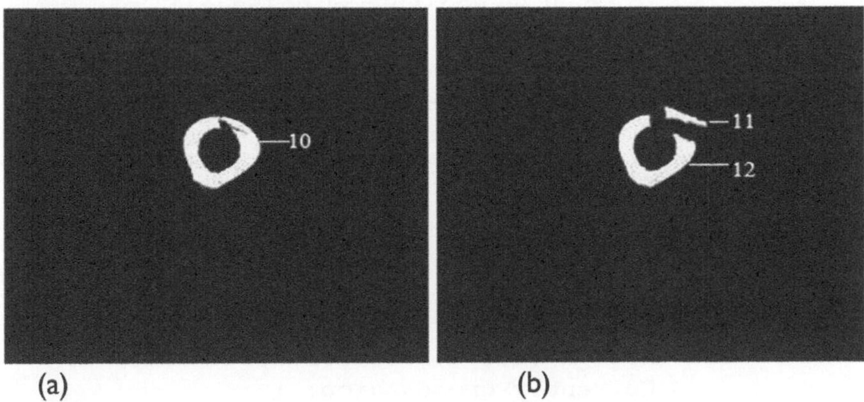

FIGURE 7.9
Labeling process: (a) ith and (b) I +1th CT slice with appropriate labels.

Algorithm 2: BoneRegionExtractionAndLabeling

```
I = read FirstSlice;
N=0; n=0; //N number of bones n number of fracture pieces per bone
while (I is not empty) do
  Seed = SeedPointSelection(I);
  Count = RegionGrowing(Seed);
    if (Count < 50) then
      Discard Seed;
    else
      N = N + 1;
        Label = concate (N,n);
        Assign lable Lable to CurrentlySpreadRegion;
        Add seed, Count and Label in GlobalList;
        I = I - CurrentlySpreadRegion;
        I_out = I_out + CurrentlySpreadRegion;
      end if
 end while
 for all slices do
   for all Seeds in GolbalList do
     Count = RegionGrowing(Seed);
     I = I - CurrentlySpreadRegion;
     Difference = GlobalList.Count - Count;
     if (Difference > 50) then
        if (Difference is positive) then
           // Introduction of new fractured piece
           Seed = SeedPointSelection(I);
           Count = RegionGrowing(Seed);
           n = n + 1;
           Label = append (GlobalList.Label, n);
           Assign lable Lable to CurrentlySpreadRegion;
           Add seed, Count and Label in GlobalList
        else
          // morphological opening to separate wrongly connected fractured piece.
 end if
     else
        Assign same label GlobalList.label to CurrentlySpreadRegion;
     end if
     I_out = I_out + CurrentlySpreadRegion;
   end for
 end for
```

7.5 Results

7.5.1 Application to Real Patient-Specific Images

In order to test the performance of the proposed method, this isused to segment and label both healthy and fractured bones from the real patient-specific CT stack. The collected CT stacks have variations in several factors, such as image resolution, bone under supervision, and fracture complexity. Still, the proposed method precisely performs the segmentation and labeling task. Figure 7.10 (a) shows a healthypatella and femur. Figure 7.10 (b) shows the resultant image. The labels represented in the current CT stack have two healthy bones without fractures. Figure 7.10 (c) shows a CT with a fracture in the tibia and fibula. Figure 7.10 (d) shows the result. The labels indicate that the image has two individual bones and each has two fractured pieces.

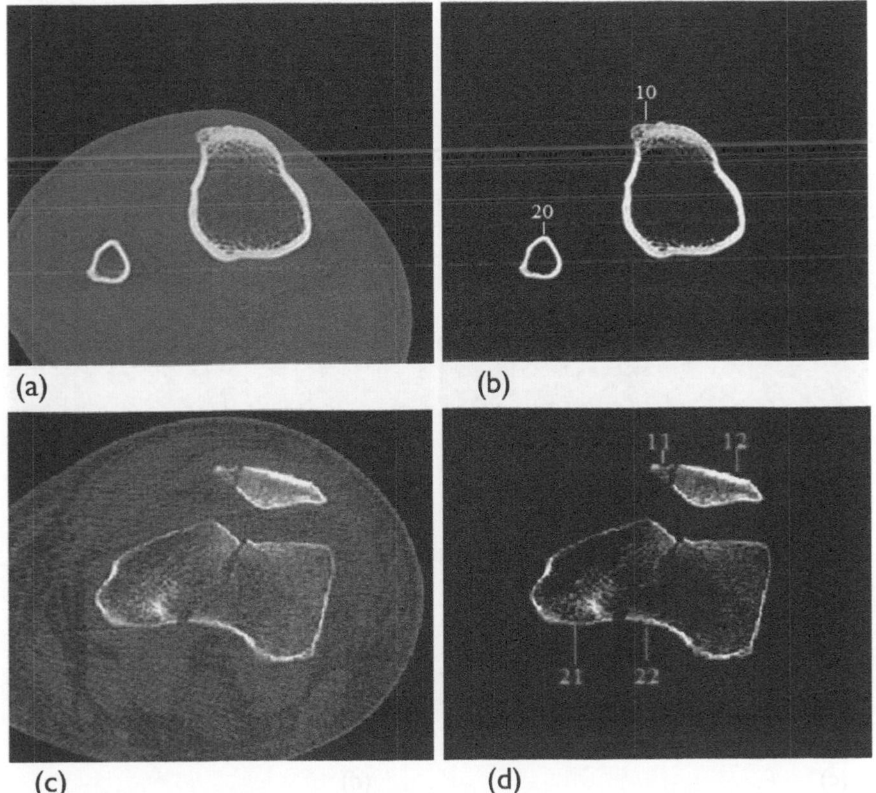

FIGURE 7.10
In the top row (a) the CT image hasa healthy bone, (b) its result; in the bottom row, (c) CT image with fractured bone, (d) its result.

7.5.2 Clinical Ground Truth

The eventual users of the CAD system are experts in the orthopedic domain. Hence, to comment on the correctness of the results, the result of the proposed method is compared with clinical ground truth obtained from experts. The union of the resultant image and the annotated image is performed. That is, the resultant image is superimposed to the clinical ground truth. Figure 7.11 (a) shows the CT image with a fracture to the femur bone, whereas Figure 7.11 (b) shows its annotated version. The result of the proposed method is shown in Figure 7.11 (c). Figure 7.11 (d) shows the union of the resultant and annotated images. The pixels highlighted in yellow color indicate the common

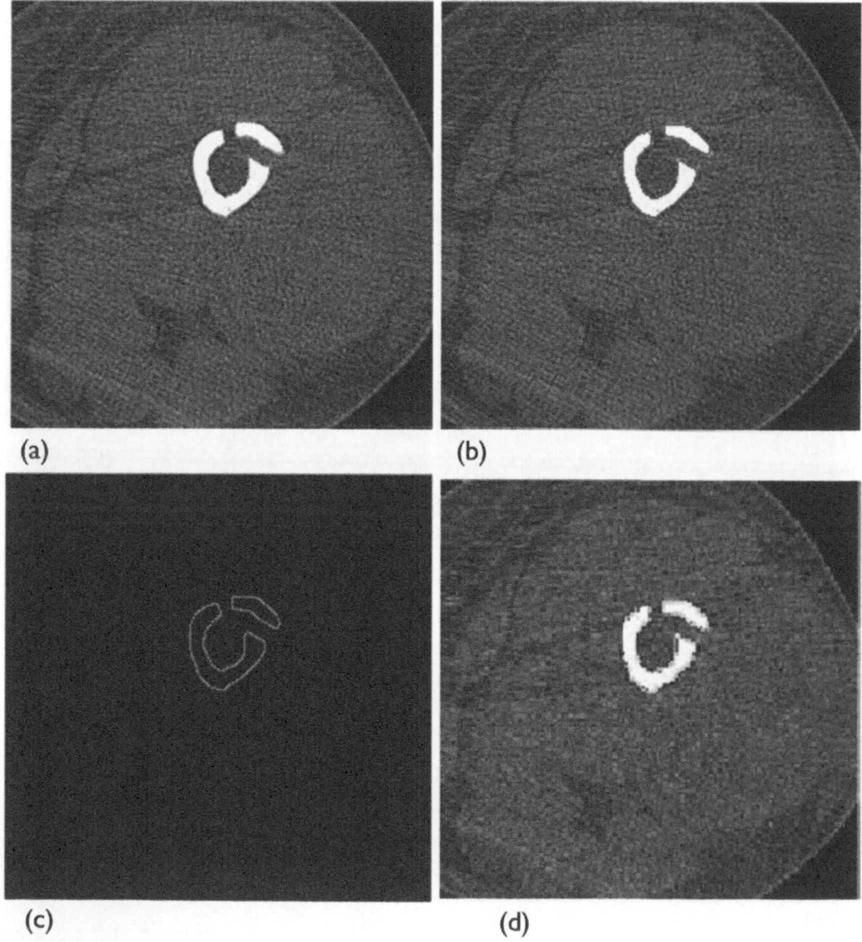

FIGURE 7.11
Comparison with clinical ground truth (a) CT image, (b) annotated CT image, (c) result of the proposed method, and (d) union set.

pixels in both the images. Highlighted pixels denote the similarity percentage. More highlighting denotes the result of the proposed method is mostly the same as that of the clinical ground truth. The experiment was conducted on 1,000 annotated images (of 8,000 CT images) and achieved 95% accuracy. Equation 7.1 is used to calculate the accuracy.

$$\text{Accuracy} = \text{Number of pixels in union set}/\text{Number of pixels in annotation set} \quad (7.2)$$

7.5.3 Comparison with State-of-the-Art Methods

To test the performance of the proposed technique, this is compared with some other segmentation techniques. Other researchers successfully implement these techniques to segment and label healthy and fractured bones. We selected thresholding [11,35], 3D region growing [17], and graph cut methods [14] for the comparison.

All the methods are used to segment fractured pieces in cortical bone tissue regions as well as in cancellous bone tissue regions. A comparison of these segmentation methods is provided in Table 7.1. The results of these segmentation methods are shown in Figure 7.12. The first column shows the results of segmentation methods applied to a CT image with a fracture in the cortical tissue region, whereas the second column shows the resultant images obtained by applying the same segmentation techniques on a CT image with a fracture in the cancellous tissue region.

The thresholding-based method is simple to implement and does not require user interaction. However, it is not a very applicable segmentation method to extract fractured pieces from the cancellous tissue region. As

TABLE 7.1

Comparison of Several Segmentation Methods

Method	Input	Output	Advantages	Limitations
Thresholding	Threshold value	Several regions	No user intervention	No labelingis performed
Graph cut	Sample points	A region per sample point	Perform labeling	Requires user intervention, and post-processing
3D region growing	Seed point	A region per seed point	Perform labeling	Does not split wrongly connected pieces
Proposed CAD model	–	Individual fractured pieces with unique labels	Auto seed point section	Requires preprocessing

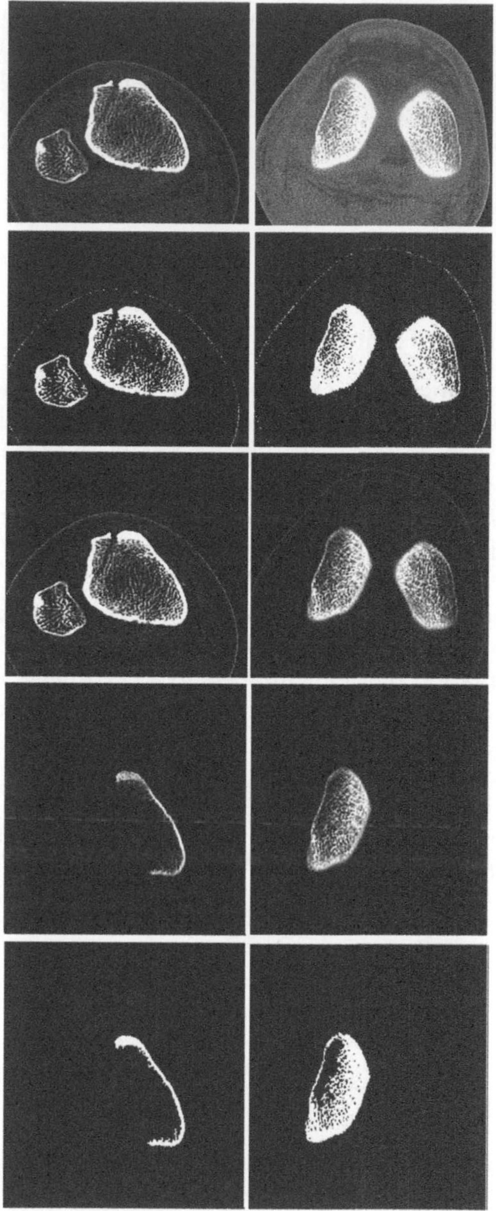

FIGURE 7.12
Resultant image obtained after the application of several segmentation methods. First-row real CT images have fractures in thecortical and cancellous tissue regions respectively. The results obtained after applying thresholding, graph cut, 3D region growing, and the proposed methodare shown in the second, third, fourth, and fifth columns respectively.

the intensity values of cancellous bone tissues and flesh are very similar, the thresholding-based segmentation method may lead to ambiguous results. Also, it does not provide unique labels to the fractured piece. The graph cut–based segmentation method separates the background and bone tissue region precisely. However, it requires a lot of user intervention in terms of sample points, and some noise remains in the image, hence post-processing is needed to remove unwanted noise. A region growing–based segmentation technique is more suitable to segment, and assigns unique labels to each fractured piece. However, the existing method in the literature requires a lot of intervention in term of seed points. In addition to this region, growing-based methods give better results after the application of efficient image enhancement methods. To remove the above-mentioned limitations of the region growing–based segmentation method, an image content–based unwanted artifacts removal method is devised in the proposed work. In addition to this, seed points are also identified automatically.

7.6 Conclusions

In this paper, a CAD system for bone fracture detection and analysis is developed. In that, 8,000 patient-specific CT images are collected from several radiology centers and hospitals in India. A histogram stretching-based preprocessing technique is devised which is responsible for removing unwanted artifacts such as flesh and to enhance the bone tissue region precisely. The devised processing technique performs image content–specific enhancement. A simple 2D region growing–based segmentation method is adopted to extract the fractured bone tissue region andto assign a unique label to each fractured piece by considering bone anatomy. In addition to this, the proposed CAD system provides several fractured features such as a count of the number of bones and number of fractured pieces per bone, which can help experts decide on an optimal recovery plan. To comment on construct validity, the experiments are conducted on real patient-specific data and results are compared with several states of the methods as well as with clinical ground truth (annotated image) provided by expert orthopedic surgeons. By confirming the result, the proposed methodology shows 95% accuracy, and it is a simple CAD model to detect and analyze bone fractures from CT images. In future, we are planning to make the database publically available (for research purpose) along with the clinical ground truth. In addition to this, we aim to develop a machine learning–based algorithm to annotate the rest of the images in the database.

Acknowledgment

The first author thanks the Ministry of Electronics and Information Technology (MeitY), New Delhi for granting the Visvesvaraya Ph.D. fellowship through file no. PhD-MLA\4(34)\201-1 Dated: 05/11/2015.

The first author would like to thank Dr. Jamma and Dr. Jagtap for providing expert guidance on bone anatomy. Along with this, he also would like thank Prism Medical Diagnostics lab, Chhatrapati Shivaji Maharaj Sarvopachar Ruganalay, and Ashwini Hospital for providing patient-specific CT images.

References

1. D. D. Ruikar, R. S. Hegadi, and K. C. Santosh, "A systematic review on orthopedic simulators for psycho-motor skill and surgical procedure training," *Journal of Medical Systems*, vol. 42, no. 9, p. 168, 2018.
2. F. Paulano, J. J. Jiménez, and R. Pulido, "3D segmentation and labeling of fractured bone from CT images," *The Visual Computer*, vol. 30, no. 6–8, pp. 939–948, 2014.
3. K. A. Egol, K. J. Koval, and J. D. Zuckerman, *Handbook of Fractures*. Lippincott Williams & Wilkins, 2010.
4. S. Montani and R. Bellazzi, "Supporting decisions in medical applications: the knowledge management perspective," *International Journal of Medical Informatics*, vol. 68, no. 1–3, pp. 79–90, 2002.
5. D. L. Pham, C. Xu, and J. L. Prince, "Current methods in medical image segmentation," *Annual Review of Biomedical Engineering*, vol. 2, no. 1, pp. 315–337, 2000.
6. G. N. Hounsfield, "Computed medical imaging," *Medical Physics*, vol. 7, no. 4, pp. 283–290, 1980.
7. T. Shapurian, P. D. Damoulis, G. M. Reiser, T. J. Griffin, and W. M. Rand, "Quantitative evaluation of bone density using the Hounsfield index," *International Journal of Oral & Maxillofacial Implants*, vol. 21, no. 2, 2006.
8. T. Velnar, G. Bunc, and L. Gradisnik, "Fractures and biomechanical characteristics of the bone," *Surgical Science*, vol. 6, no. 06, p. 255, 2015.
9. F. B. Sachse, "5digital image processing," In *Computational Cardiology*, 2004, pp. 91–118, Springer, Berlin, Heidelberg, Germany.
10. S. Vasilache and K. Najarian, "Automated bone segmentation from pelvic CT images," in *Proceedings of the IEEE Workshop on Bioinformatics and Biomedicine (BIBMW)*, 2008, pp. 41–47.
11. M. Tomazevic, D. Kreuh, A. Kristan, V. Puketa, and M. Cimerman, "Preoperative planning program tool in treatment of articular fractures: process of segmentation procedure," in *XII Mediterranean Conference on Medical and Biological Engineering and Computing*, 2010, pp. 430–433, Springer, Berlin, Heidelberg.
12. S. Tassani, G. K. Matsopoulos, and F. Baruffaldi, "3D identification of trabecular bone fracture zone using an automatic image registration scheme: a validation study," *Journal of Biomechanics*, vol. 45, no. 11, pp. 2035–2040, 2012.

13. Y. He, C. Shi, J. Liu, and D. Shi, "A segmentation algorithm of the cortex bone and trabecular bone in proximal humerus based on CT images," in *Automation and Computing (ICAC)*, 201723rd International Conference on IEEE, 2017, pp. 1–4.
14. J. Fornaro, G. Székely, and M. Harders, "Semi-automatic segmentation of fractured pelvic bones for surgical planning," in *International Symposium on Biomedical Simulation*, 2010, pp. 82–89, Springer, Berlin, Heidelberg, Germany.
15. M. Harders, A. Barlit, C. Gerber, J. Hodler, and G. Székely, "An optimized surgical planning environment for complex proximal humerus fractures," in *MICCAI Workshop on Interaction in Medical Image Analysis and Visualization* (Vol. 30), 2007.
16. J. Kaminsky, P. Klinge, T. Rodt, M. Bokemeyer, W. Luedemann, and M. Samii, "Specially adapted interactive tools for an improved 3D-segmentation of the spine," *Computerized Medical Imaging and Graphics*, vol. 28, no. 3, pp. 119–127, 2004.
17. R. K. Justice, E. M. Stokely, J. S. Strobel, R. E. Ideker, and W. M. Smith, "Medical image segmentation using 3D seeded region growing," In *Medical Imaging 1997: Image Processing* (Vol. 3034), 1997, pp. 900–911, International Society for Optics and Photonics.
18. P. Y. Lee, J. Y. Lai, Y. S. Hu, C. Y. Huang, Y. C. Tsai, and W. D. Ueng, "Virtual 3D planning of pelvic fracture reduction and implant placement," *Biomedical Engineering: Applications, Basis and Communications*, vol. 24, no. 03, pp. 245–262, 2012.
19. T. B. Sebastian, H. Tek, J. J. Crisco, S. W. Wolfe, and B. B. Kimia, "Segmentation of carpal bones from 3D CT images using skeletally coupled deformable models," in *International Conference on Medical Image Computing and Computer-Assisted Intervention*, 1998, pp. 1184–1194, Springer, Berlin, Heidelberg, Germany.
20. T. Gangwar, J. Calder, T. Takahashi, J. E. Bechtold, and D. Schillinger, "Robust variational segmentation of 3D bone CT data with thin cartilage interfaces," *Medical Image Analysis*, vol. 47, pp. 95–110, 2018.
21. W. Shadid and A. Willis, "Bone fragment segmentation from 3D CT imagery using the probabilistic watershed transform," in *Southeastcon, 2013 Proceedings of IEEE*, IEEE, 2013, pp. 1–8.
22. W. G. Shadid and A. Willis, "Bone fragment segmentation from 3D CT imagery," *Computerized Medical Imaging and Graphics*, vol. 66, pp. 14–27, 2018.
23. M. H. Moghari and P. Abolmaesumi, "Global registration of multiple bone fragments using statistical atlas models: feasibility experiments," in *Engineering in Medicine and Biology Society, 2008*. EMBS 2008. 30th Annual International Conference of the IEEE, IEEE, 2008, pp. 5374–5377.
24. K. C. Santosh, L. Wendling, S. Antani, and G. R. Thoma, "Overlaid arrow detection for labeling regions of interest in biomedical images," *IEEE Intelligent Systems*, vol. 31, no. 3, pp. 66–75, 2016.
25. P. Szymor, M. Kozakiewicz, and R. Olszewski, "Accuracy of open-source software segmentation and paper-based printed three-dimensional models," *Journal of Cranio-Maxillofacial Surgery*, vol. 44, no. 2, pp. 202–209, 2016.
26. M. L. Poleti, T. M. F. Fernandes, O. Pagin, M. R. Moretti, and I. R. F. Rubira-Bullen, "Analysis of linear measurements on 3D surface models using CBCT data segmentation obtained by automatic standard pre-set thresholds in two segmentation software programs: an in vitro study," *Clinical Oral Investigations*, vol. 20, no. 1, pp. 179–185, 2016.

27. S. G. Armato III, G. McLennan, M. F. McNitt-Gray, C. R. Meyer, D. Yankelevitz, D. R. Aberle, C. I. Henschke, E. A. Hoffman, E. A. Kazerooni, H. MacMahon, and A. P. Reeves, "Lung image database consortium: developing a resource for the medical imaging research community," *Radiology*, vol. 232, no. 3, pp. 739–748, 2004.
28. D. Testi, P. Quadrani, and M. Viceconti, "PhysiomeSpace: digital library service for biomedical data," *Philosophical Transactions of the Royal Society of London A: Mathematical, Physical and Engineering Sciences*, vol. 368, no. 1921, pp. 2853–2861, 2010.
29. D. D. Ruikar, R. S. Hegadi, and K. C. Santosh, "Contrast stretching-based unwanted artifacts removal from CT images in recent trends in image processing and pattern recognition (accepted)," Springer, 2019.
30. K. C. Santosh, S. Candemir, S. Jäger, L. Folio, A. Karargyris, S. Antani, and G. Thoma, "Rotation detection in chest radiographs based on generalized line histogram of rib-orientations,"in *Computer-Based Medical Systems (CBMS), 2014 IEEE 27th International Symposium on IEEE*, 2014, pp. 138–142.
31. R. C. Gonzalez and R. E. Woods, *Digital Image Processing* (2nd ed.) Publishing House of Electronics Industry, Beijing, China, 2002, p. 455.
32. S. Vajda and K. C. Santosh, "A fast k-nearest neighbor classifier using unsupervised clustering," in *International Conference on Recent Trends in Image Processing and Pattern Recognition*, Springer, Singapore, 2016, pp. 185–193.
33. A. Karargyris, J. Siegelman, D. Tzortzis, S. Jaeger, S. Candemir, Z. Xue, K. C. Santosh, S. Vajda, S. Antani, L. Folio, and G. R. Thoma, "Combination of texture and shape features to detect pulmonary abnormalities in digital chest X-rays," *International Journal of Computer Assisted Radiology and Surgery*, vol. 11, no. 1, 2016, pp. 99–106.
34. R. Adams and L. Bischof, "Seeded region growing," *IEEE Transactions on Pattern Analysis and Machine Intelligence*, vol. 16, no. 6, pp. 641–647, 1994.
35. J. Y. Lai, T. Essomba, and P. Y. Lee, "Algorithm for segmentation and reduction of fractured bones in computer-aided preoperative surgery," in *Proceedings of the 3rd International Conference on Biomedical and Bioinformatics Engineering*, ACM, 2016, pp. 12–18.

8

A Systematic Review of 3D Imaging in Biomedical Applications

Darshan D. Ruikar, Dattatray D. Sawat, K.C. Santosh, and Ravindra S. Hegadi

CONTENTS

8.1 Introduction

Scientific visualization is a method of scientific computing. It converts acquired symbolic data into a geometric form to convey silent information of underlying data and to see the unseen structure which is beneficial for comprehension, analysis, and interpretation [1, 2]. The field of scientific visualization has been widely explored in the last three decades. The National Science Foundation Visualization created the field of scientific visualization in 1987 by presenting a paper in a scientific computing workshop [3]. The phenomenal growth in scientific visualization is possible due to advances in data acquisition and computational technologies. The acquired data can be two-, three-, or even more dimensional, and is used to convey detailed information about complex phenomena/processes such as the flow of gases and fluids, biological processes, space, and earth sciences. The acquired data are voluminous and exclusively in numerical format; it is impossible for the human brain to analyze and interpret thosedata. Most of the valuable information may be lost in the manual process. Thus scientists are motivated to adopt technological advancements [4].

Nowadays, scientific visualization is effectively used to simulate physical processes in order to achieve a better and precise understanding of our universe. This is used to study natural phenomena which are extremely large or small, unduly quick or slow, or might be too harmful or dangerous to observe directly. In addition to this, scientific visualization helps researchers to extract meaningful information as well as to discover new information from complex and voluminous datasets by using interactive computer graphics and imaging techniques. Scientific visualization is the basis behind the growth of several fields, such as computer graphics, image processing, computer vision, signal processing, cognitive sciences, computational geometry, user interfaces, and computer-aided design. Moreover, visualization is the backbone of a wide variety of applications such as defense (computer-generated forces and advanced distributed simulation applications, for instance), engineering (computer-aided design and computer architecture for instance), computational fluid dynamics, and computer graphics applications [1].

In addition to this, visualization plays a vital role in most computerized medical applications such as computer-aided diagnosis (CAD), computer-assisted surgery (CAS), and simulator development. In CAD applications, visualization is essential for accurate disease diagnosis and prognosis. For instance, it is possible to visualize the actual anatomical structure of bones or several soft tissues, and their responses for particular situations. Moreover, experts can identify the effects of procedures before treating patients. CAS visualization plays a crucial role in the development of custom prostheses and anatomic models. Other than this, visualization-based CAS systems are successfully developed for neurosurgery, image-guided surgery, custom anatomic atlas, robotic assistance, and surgical planning [1]. Virtual reality (VR)-based simulator development is an emerging field in medicine, where (3D) visualization plays a vital role. Most of the VR-based simulators are developed to increase surgical competence and reduce runtime complications. Especially in the orthopedic field, such types of VR-based trainer simulators are in demand. With the help of simulators novice trainees can do the practice of various orthopedic psychomotor skills at no cost once it has been developed [5]. In addition to this, simulators are helpful for 3D model visualization of complex bone anatomy.

This chapter aims to provide detailed information on several volume visualization principles, techniques, and algorithms that are widely used in the medical field. Explanations on volume visualization techniques used in the rest of the fields discussed above are beyond the scope of this chapter. Section 8.2 discusses some preliminaries of volume visualization. It includes the explanation of volumetric data and their types, several grid structures, and the data acquisition process. Sections 8.3 and 8.4 respectively cover the detailed information on the various indirect and direct volume rendering techniques used in the literature to render medical data. Section 8.5 discusses the challenges in volume rendering on conventional computational devices and explores the recent advances in volume visualization. The primary focus is given toexploring advances in hardware-based volume visualization and transfer functions (TFs). This section also covers the advantages of GPU utilization in both direct and indirect volume rendering techniques. Section 8.6 liststhe various commonly used tools and libraries for volume visualization. Section 8.7 provides the conclusion and future directions.

8.2 Volumetric Data

Before proceeding to a detailed explanation of several volume visualization approaches in the biomedical field, it is necessary to explore some basic concepts that are required to understand the logic of visualization algorithms. This section covers the required basic terminologies and various data acquisition alternatives available in the medical field to acquire volume data. Different ways of representing volume data, various grid (mesh) structures, and general steps in the volume visualization process are explained in detail in this section.

8.2.1 Data Acquisition

Volume datasets can be generated by using sampling techniques (stochastic method for instance), simulation, modeling techniques, curving the material, or hand painting in 3D. Other than these, voxelizing geometric description of objects and writing programs are the most popular alternatives to generate volume datasets [6]. However, in the medical field, volume datasets are often acquired by scanning the material-of-interest using 3D scanning techniques such as magnetic resonance imaging (MRI), computer-aided tomography (CT), positronemission tomography (PET), and/or sonogram machines. In addition to this, laser scan confocal and other high-power microscopes are also used to acquire data [7]. Depending upon the concern of inspection, different 3D scanning techniques are used. For instance, to inspect soft tissues, MRI is preferable, whereas for inspecting bone diseases and trauma CT is widely used.

Due to the evolution of medical imaging modalities (such as CT, MRI, and PET for instance) after 1970 the terminologies in medical imaging have shifted from 2D projection to fully isotropic 3D images [8]. Commonly, these medical imaging modalities scan material-of-interest from several angles and generate anumber of 2D slices. The count of slice per stack is highly dependent upon the amount of portion one needs to scan, the thickness of each slice, and the distance between two slices. The radiographer sets these parameters during the time of inspection. Then the sequence of 2D slices obtained from the scanner is used to reconstruct the volumetric model in 3D space. The 3D reconstructed models can be used for better visualization of internal structure, for diagnosis or to plan the recovery process.

8.2.2 Volume Data

Volume data is a set of samples (x, y, z, v). The values of v represent the quantitative property of scanning material at a 3D location (x, y, z). The values of v can be scalar or vector. More generally, scalar values are single-valued, and represent some measurable property of the data such as color, intensity, heat, or pressure. However, the vector value is multi-valued, and represents the velocity or direction in addition to the scalar value at that location [6]. Vector values are useful in the fields where direction plays a vital role during volume visualization and interpretation process (computational fluid dynamics for instance). A further discussion of vector volume data is out of the scope of our study, because this paper aims to provide extensive information about volume visualization techniques thatare useful in the biomedical field. In this field, the volume data generated with the help of 3D scanning devices by scanning material of interest is scalar data. The value of v represents the gray intensity values at that location.

In the biomedical field, volume is a 3D array of volume-elements (voxels). This 3D array can be visualized as a stack of 2D slices as shown in Figure 8.1. Slice-oriented representation is another way to visualize 3D data. This is the traditional representation, nothing but how the physicians look at volume dataset. It is denoted by a matrix $V = \Gamma^{X \times Y \times Z}$ where X, Y, and Z represent rows, columns, and slices respectively. The V is a discrete grid of voxels v. Each voxel v is represented by $I(v): N^3 \rightarrow \Gamma$ [9]. Concerning CT imaging, it is

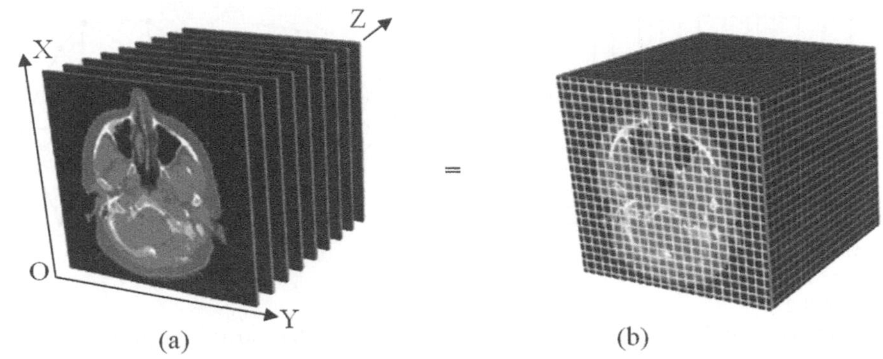

FIGURE 8.1
3D volume data representation (a) n × 2D slices and (b) 3D stack.

gray values which are X-ray attenuation coefficient of the material of interest at that point. The volume data obtained from CTscanners are anisotropic with an equal sampling density in the x- and y-directions (more generally, 512 × 512 voxels in each direction), whereas it has a coarser density along the z-direction (i.e. the number of slices may vary per stack). The number of slices per stack ranges from 100 to 600, and typically depends upon area under supervision and severity of disease/trauma. The datasets V generated by the CT-scanning process arethe basis for the development and analysis of volume visualization (rendering) algorithms.

8.2.3 Grid Structures

Grid (mesh) structures determine the volume visualization technique. They structures depend on the source of volumetric data. A grid structure can be structured or unstructured [10]. A structured grid shows regular connectivity between the grid points, whereas an unstructured grid shows intermittent connectivity. Figure 8.2 shows the different types of grid structures. Uniform, rectilinear, and curvilinear are examples of structured grids as shown in Figure 8.2 (a), (b), and (c) respectively. Unstructured grids (Figure 8.2 (d)) are usually generated by physical simulation, whereas scanning devices produce volumetric data in structured grids. The grids generated by CT or MRI scanners are a rectilinear structured grid. Further discussion on rest of the grid structure is beyond the scope of this paper.

8.2.4 Volume Visualization

Volume visualization techniques are used to create 2D graphical representations from volume datasets defined over 3D grids. Based on the types of volume datasets, different techniques can be adopted for volume visualization (rendering) [11]. Techniques such as isosurfaces, slice planes, and contour

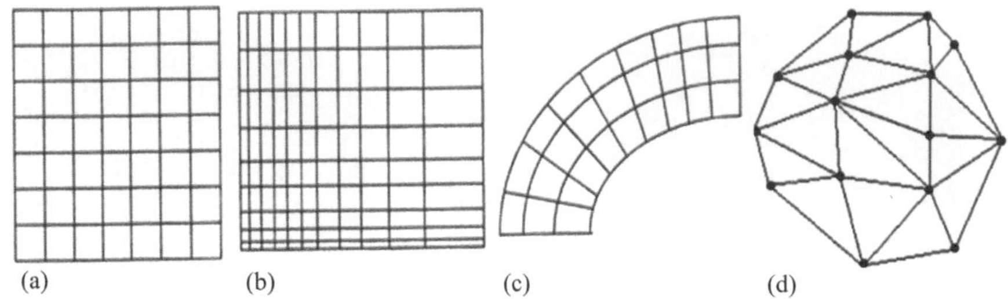

FIGURE 8.2
Types of grid (a) uniform, (b) rectilinear, (c) curvilinear, and (d) unstructured grid.

slices are best suited for the visualization of scalar data, whereas streamlines, cone plots, and arrow plots are suitable visualization techniques for vector data. More explanation on visualization techniques of vector data is out of the scope of this paper, as medical 3D scanning devices generate scalar data. Further, scalar data volume visualization techniques can be categorized into parts: direct and indirect volume visualization. Sections 8.3 and 8.4 provide a detailed explanation of these two techniques.

8.2.5 Steps in Volume Visualization

Most of the volume visualization methods use common steps to render volumetric data. Data acquisition, preprocessing, and define view are some common steps included in many state of the art volume visualization methods [7].

8.2.5.1 Data Acquisition and Dimension Reconstruction

The initial step of every visualization method is data acquisition. The material of interest is scanned via scanning devices to generate the data. In the biomedical field, CT or MRI scanners are commonly used to generate the volume data. The acquiesced data need to be reconstructed to match the dimensions of the scanned material of interest to the dimensions of the display coordinate system.

In addition to this, reconstruction is required to make equidistant (regularly spaced) scanned data. This is required when aradiographer intentionally constructs the irregular spaced data to capture a complicated part in detail. The radiographer may increase the distance between the adjacent slices where the scanned portion is less interesting and may decrease the distance at the area of interest. For example, during CT stack construction, the radiographer may increase the distance while scanning healthy bone and decrease the distance during the scanning of the fracture-prone area to obtainmore details [23]. In such a situation, new slices need to be reconstructed or to replicate existing slices to make the volume equidistant. Most of the methods adopted the interpolation method to predict the values in new slices.

8.2.5.2 Data Preprocessing and Extraction

The acquired slices may contain unwanted artifacts (CT bed, cables, and flesh, for instance), or noise may get added during the image acquisition process. At this point, suitable preprocessing and segmentation techniques need to be applied. The preprocessing technique is responsible for removing unwanted artifacts and forenhancing the desired portion with precision [74]. After that, the application of a segmentation technique necessary to extract the desired portion out of the image. Some authors named this as "data classification step."

8.2.5.3 View Definition

The main aim of the view definition step is the selection of an appropriate TF. The TF is responsible for mapping volume data onto geometric or display primitives. In addition to this, it specifies the coloring and lighting effects. Lighting effects are used to enhance the visibility of the surface shape and to provide the 3D perspective of the volume data. Next, the view definition includes adjusting camera position, specifying aspect ratio, and selecting-projection type. This step may vary in every algorithm. During rendering, these primitives can be stored, manipulated, and intermixed with each other to display the view on the screen.

8.3 Indirect Volume Rendering (Surface Fitting)

Indirect volume (i.e. surface) rendering techniques render the only surface of the given volume data. The rendered surface is opaque, and it is easily manageable. The coherent structures like skin and bone are represented by point sets with the same sampling rate. Polygons are used to approximate the surface from the point set. Then the triangular mesh is created to better represent the given volume, and at last the created 2D triangular mesh is rendered on screen with proper shading and lighting effects [6].

The surface rendering approach commonly extracts geometric primitives (the boundary of an object, for instance) to approximate the surface of objects being rendered [9]. To extract such geometric primitives, a suitable segmentation technique is adopted. Most surface rendering methods use a thresholding-based segmentation technique to extract the boundaries of the desired objects. This is true even for commercial medical imaging tools such as DICOM* viewer and 3D slicer, for instance. They also use a user-defined

* DICOM: digital imaging and communications in medicine.

threshold to extract boundaries of objects before generation of polygon mesh for surface rendering.

In short, the surface rendering approach segregates each voxel in one of two classes: one part of the desired object and another part of the background. To do this, it needs the help of a user-defined threshold value. Then, object boundaries are extracted using an edge detector operator (Canny or Sobel operator, for instance). At last, a polygonal mesh is generated and the surface is displayed on screen by applying proper shading effects. Shading effects are determined by the isosurfaces value at that location [9]. Opaque cubes (cuberilles), contour tracing, marching cubes/tetrahedral, and dividing cubes are afew examples of surface rending algorithms. Amongst them, marching cube is a widely used method.

8.3.1 Opaque Cubes (Cuberilles)

This method initially performs the binarization of a given volume by considering the isovalue. Then all boundary front faces are identified such that the normal points of the faces are pointing towards the viewpoint. Lastly, these faces are rendered as shaded polygons. This method does not use any interpolation method to determine points, thus the object boundary is not identified precisely; moreover, it results in a block surface.

8.3.2 Contour Tracing

This is one of the most flexible methods in the biomedical field for volume visualization before the invention of the marching cube algorithm. Like cuberille, this method also affects binarization of the given volume. Then polyline is obtained by traversing the boundary pixels in a clockwise direction. Finally, the polylines of adjacent slices thatrepresent the same objects are connected to form triangles, and the generated triangles are displayed on the screen as an isosurface of the object. This method becomes stuck if there is considerable variation between two adjacent slices and there are some objects in the slice.

8.3.3 Marching Cube

Lorensen et al. [31] devised the marching cubes algorithm to determine the isovalued surface with a triangle mesh. The algorithm defines a voxel in terms of a cube having pixels values at eight corners of the cube. These cubes are marched through the entire volume. During marching, each vertex of the cube is classified as being inside or outside the isosurface. The edges where one vertex is classified as outside and another one is classified as inside are used to form a triangular patch. Then these triangular patches from adjacent lines are connected to each other to form the isosurface. To form a triangular patch between two adjacent slices, the cube (voxel) can be represented by 256 different cases. These are reduced to only 15 cases observing symmetrical structures. These 15 generic triangles are stored into a lookup table for future reference. The actual vertices to form a triangle are determined by the linear interpolation function.

Lastly, the normal value is determined for each vertex, and the triangular mesh is projected on standard graphics hardware for rendering. Due to simplicity and efficiency, the marching cube algorithm is adopted by several researchers to reconstruct 3D models from acquired medical data [12]. The divide cubes are used to render the isosurfaces in large datasets, whereas the variation of marching cubes, the marching tetrahedra [32], is used to render the surface in unstructured grids.

The surface rendering technique best approximates the surface of the volume data, andis efficient and straightforward concerning space and time. However, in addition to surface most of the applications require internal details also. For instance, in VR-based orthopedic simulators developed to compute drilling parameters such as depth, an entire volume is required to perform erosion. That is to remove the part of volume drilled by the drill bit to show the depth. In such applications' surfaces, rendering techniques are not efficient. Hence, many such applications havenow adopted the direct volume rendering technique.

8.4 Direct Volume Rendering

Surface rendering techniques extract geometric primitives to display the surface of the volume data. If the data size is too large, then the time required to extract geometric features and render a surface may be very long. In such cases, surface rendering techniques are not efficient. In addition to this, the volume is represented by the surface, thus most of the internal information is lost during the rendering process. That information is valuable and required for a precise understanding of the data. To avoid information loss andto gain high accuracy, direct volume rendering (DVR) techniques are being used frequently in biomedical applications [6].

DVR techniques directly map voxels to pixels in a 2D plane. That is, they directly render the segmented volume with the help of a suitable TF without extracting any kind of geometric features or structures. The DVR approach reconstructs the 2D image directly from 3D volume data. Raycasting, splatting, shear-wrap, maximum intensity projection, and 3D texture mapping are the most commonly used DVR techniques. These techniques operate on actual volume data.

8.4.1 Raycasting

Raycasting is one of the most widely used DVR techniques to display volume data in the 2D viewing plane. It gained a lot of importance in the literature

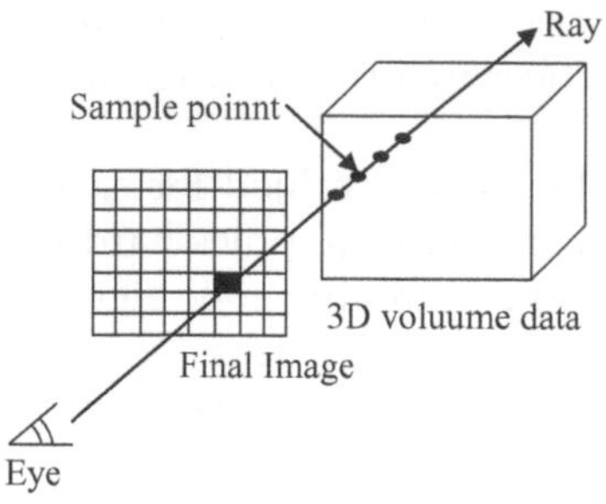

FIGURE 8.3
Raycasting pipeline.

as it has the most substantial body of publications. This technique fires a ray from each pixel in the viewing plane into the volume to determine the color and opacity values from each scalar value. The tri-linear interpolation function is used to compute these values [8, 9]. This technique does not generate any shadows or reflection effects [7]. The raycasting pipeline is shown in Figure 8.3. A few researchers worked to improve the performance of the raycasting method. Levoy [13] developed an adaptive refinement model. The method skips the empty spaces during raycasting and sample points computation to improve the performance.

Further, Levoy [14] discussed two improvements. In the first technique, spatial coherence in the volume data is encoded through a pyramid of binary volume; the second method uses the opacity threshold to terminate the ray tracing. A thresholding level parallelism and a bricked volume layout are used by Grimm et al. [15] to improve the performance of the raycasting method.

8.4.2 Splatting

The splatting DVR technique represents each voxel by 3D reconstruction (Gaussian) kernel. Each projected kernel leaves a splat (footprint) on the viewing plane [6, 7]. The voxel contribution to each pixel in the image plane is calculated using a look table. In addition to this, the color and opacity values in the pixels are computed by TF [8]. Figure 8.4 shows the splatting pipeline. This technique is named splatting because the rendering output is similar to a scenario generated on a glass plate when throwing a snowball on it. The concentration of snow is more at the center, whereas it fades while moving away from the center. The basic splatting algorithm suffers from a color blending problem [16]. An aligned sheet buffer technique is adopted by Westor to solve this problem [17]. A self-sufficient data structure, "FreeVoxels," was developed [18] to remove the same problem. Zwicker et al. [19] used an elliptical weighted average filter to overcome the color blending problem. Xeu et al. [20] adopted texture mapping to improve the performance of the splatting process.

8.4.3 Shear-Warp

A hybrid DVR algorithm shear-warp is one of the fastest volume rendering algorithms. It attempts to combine the advantages of both the image order-based and object order-based volume rendering methods. It uses run-length encoding (RLE) compression to compress volume the data, which allows fast streaming through it. In addition to this, it factorizes the viewing transformation into 3D shear and warp transformation in the 2D plane [8, 9]. Figure 8.5 shows the transformation of shared object space from the parallel projection tothe original object space. The shear-warp algorithm achieves greater rendering speed at the cost of image quality. To overcome this drawback, i.e. to maintain image quality while rendering the pre-integrated volume, rendering in the shear-warp algorithm for parallel projection is implemented in Schulze et al. [21]. Further, Kye et al. [22] presented two methods to improve image quality. In the first method, super-sampling is performed in an intermediate image space, whereas the second method uses a pre-integrated rendering technique with the help of a new data structure named "overlapped min-max block."

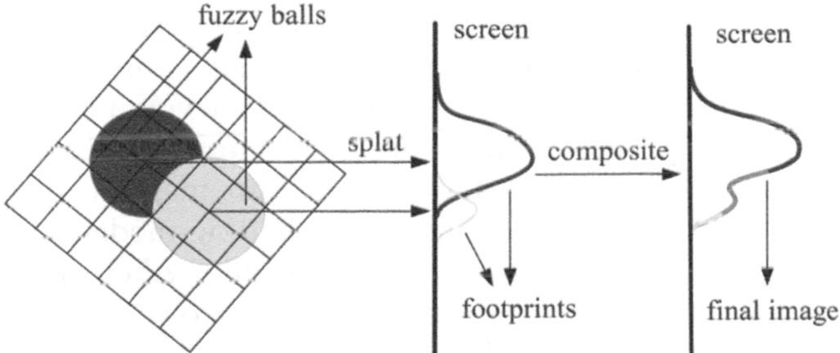

FIGURE 8.4
Splatting pipeline. (Image courtesy of Zhang et al. [8].)

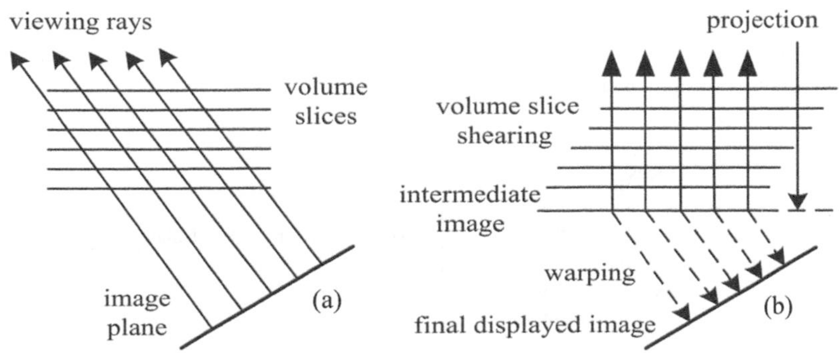

FIGURE 8.5
(a) standard transformation and (b) shear-warp factorized transformation for a parallel projection. (Image courtesy of Zhang et al. [8].)

8.4.4 Maximum Intensity Projection

The basic idea of maximum intensity projection (MIP) is to assign a maximum intensity to the pixels. The maximum intensity is determined by evaluating each voxel of volume thatlies on the path of ray coming from the viewer's eye [22]. The idea of MIP is illustrated in Figure 8.6. It is mainly used to visualize high-intensity structures such as blood vessels from the volume data. The limitations of MIP techniques are as follows: no shading information is given, and the depth and occlusion information is lost [9]. To overcome one of the limitations (depth information loss) of MIP, tri-linear interpolation—a function-based, interactive, high-quality MIP method—is implemented in Mroz et al. [24].

8.4.5 3D Texture Mapping Volume

The texture mapping volume rendering technique is widely supported by traditional display devices to render synthetic images. Thanks to Cullip et al. [25] and Cabral et al. [26], the texture mapping technique has become popular. The core idea of the texture mapping method is to decompose the volume into three stacks and interpret each voxel as 3D texture defined over $[0, 1]^3$. During rasterization, the texture information at an arbitrary point is extracted by using tri-linear interpolation within the volume data. Figure 8.7 shows the pipeline of the texture mapping volume rendering technique. Van Gelder et al. [27] introduced shading in texture mapping to improve the image quality. Further diffuse and specular shading models are integrated into Rezk-Salama et al. [28] to improve image quality. Abellán et al. [29] proposed three types of shading for the multiple-model dataset to accelerate the performance of the texture mapping technique.

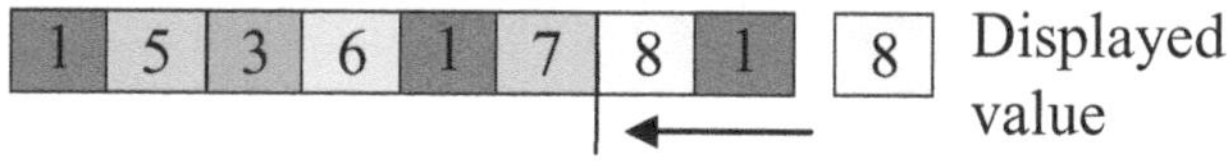

FIGURE 8.6
Concept of maximum intensity projection.

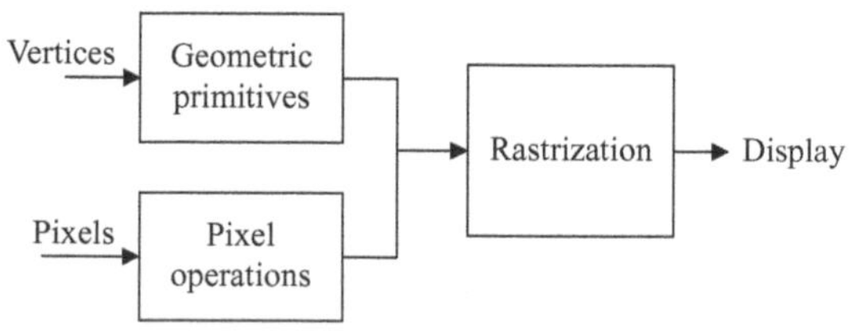

FIGURE 8.7
3D texture mapping pipeline.

8.5 Recent Advances in Volume Visualization

Visualization data arestored in multi-dimensional volumes. These volumes may grow large in spatial terms as well as in a number of dimensions. The data architecture used to store volumetric data supports various file formats (such as DICOM and STL*, for instance) and software (several DICOM viewers or 3D slicers, for instance). Devising and developing an efficient volume rendering algorithm is still a challenging task. Visualization hardware, TFs, and rendering styles are a few of the primary challenges to develop such algorithms. In addition to this, researchers need to focus on some additional challenges, such as view-dependent algorithms, image-based rendering, multi-resolution techniques, importance-based methods, adaptive resource-aware algorithms, remote and collaborative visualization algorithms, and networking [11].

8.5.1 Advances in Hardware (GPU)-Based Volume Rendering

In traditional volume rendering algorithms, the software calculates the frame, whereas the CPU is used only to render the frames. There are certain disadvantages associated with software-based volume rendering. It assumes the hardware device (CPU and memory, for instance) is fast enough. In addition to this, algorithms forcefully ask the hardware to handle data which is high in volume. However, CPU and memory by themselves cannot handle such voluminous data (dedicated hardware is required to handle such data). The reason behind this is the gap present between the speed of memory and CPU. Software-based rendering algorithms ask for precalculated data, because precalculated data are always faster than the data calculated on the fly (i.e. during the rendering process). Though the processing of precalculated data makes things faster, the process still does not meet to the desired efficiency, because memory is not fast enough to feed the data that is required by CPU. For example, the clock speed of memory (DDR3 RAM for instance) is ~1333 MHz, whereas the clock speed of the processor is ~2.6 GHz. There is a massive gap between processing speeds. Additionally, cache memories are too small.

Due to the limitations mentioned above relating to software-based volume rendering algorithms, several recent studies introduced innovative hardware-based approaches that are better suited for parallel processing. In addition to this, various advances are being made in parallel processing hardware, i.e. graphical processing unit (GPU). Hence nowadays the GPUs are at the core of volume visualization.

* STL: surface template library.

8.5.1.1 The Need for GPU

Generating volumetric data and rendering them on large size displays is always computationally expensive. To render the data in inner slices, the algorithms have to create pixels for each of the slices based on direct or indirect volume rendering techniques. Processing such large data is always challenging for the developers, and it is also a hardware-bound task. In order to explore the insights of the data, the user has to view the data from different point of views. This, in turn, creates a demand for view-dependent processing. All such processing is computationally expensive in terms of speed and memory. This leads to the adaptation of GPU for volume visualization [67].

The data processing in visualization is required at several stages such as data acquisition, data preprocessing, volume generation, volume rendering, and visualization. This has to be done in real time, or sometimes on the fly. To process such high-dimension data, a parallel processing system becomes an essential requirement. In addition to this, the pixels/voxels need to be created using volume generation algorithms for direct volume rendering. The direct volume rendering method requires the abstract and underlying data together. That is, it requires the volume to be transparent and should display the propagation of light and shadows through the volume. Such rendering makes it necessary for itto be processed by parallel systems. The required parallel processing can be achieved by GPUs [72]. In addition to this, some applications require for the volume data to be explored with insights. To explore such insights with minute details, the display screen has to be large, and the resolution of the screen must be high enough.

Other than this, the advancements in GPU are far better than CPU as far as processing speed and memory capacity are concerned. These advancements help GPUs to render frames at a very high rate. This is possible because the large size data can also be fitted easily into the GPU [72]. This property of GPU is beneficial for hardware-based volume rendering. The visualization in the medical field is increasingly dependent on high-performance computing (HPC) to calculate underlying pixel data, volumes properties, and rendering phenomena [74]. For this kind of tasks, multi-GPU workstations, i.e. distributed GPU clusters and cloud-based GPU, are a more feasible option. GPU units are multicore, and each GPU unit has multi-threaded blocks. These are the core of parallel processing. The introduction of high-bandwidth networking is the backbone for distributed image processing and is being actively applied in the field of medical visualization. Simulation-based hardware and software areincreasingly applied in visualization to generate volumes. In addition to GPU, some recent applications are making use of tensor processing units (TPU) for medical visualization tasks.

8.5.1.2 Accelerators on the GPU

To accelerate parallel processing with GPU, there is a need for software accelerators to leverage the parallel processing power of GPUs. One such

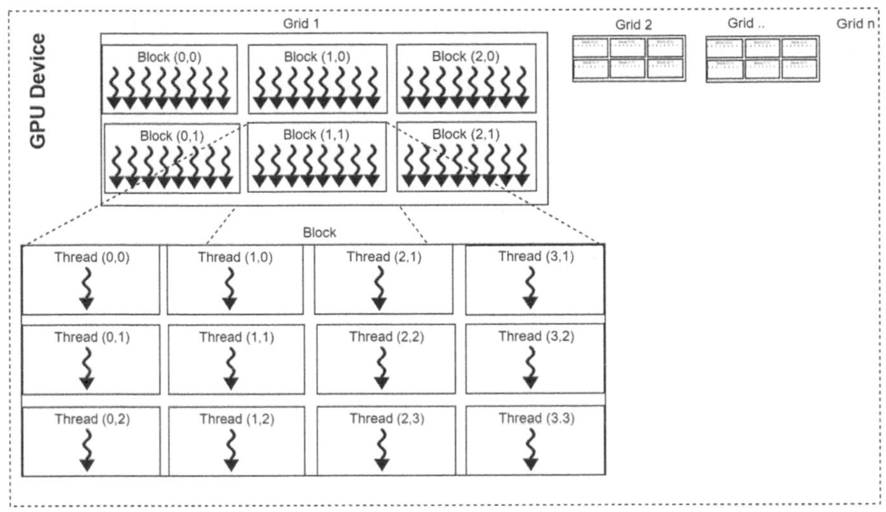

FIGURE 8.8
CUDA architecture. (Image courtesy of NVIDIA [72].)

accelerator is the compute unified data architecture (CUDA). Other than CUDA, OpenCL* is another accelerator used to improve the visualization. The GPU manufacturer AMD† supports OpenCL acceleration, and this is being used in many visualization devices for processing multi-dimensional visualization data. In addition to this, these devices are relatively inexpensive. Thus nowadays many medical visualization applications use GPU-accelerated devices [74]. However, visualization quality and execution time are dependent on factors such as GPU models, number of GPUs, number of threads, and acceleration libraries used [72].

NVIDIA introduced the CUDA model in 2006. This is used to process instructions in parallel. Most of the programming features are an extended form of C programming language, which is compiled using an NVCC compiler. The CUDA programming model uses the acceleration in GPU by executing instructions in parallel threads. Figure 8.8 shows the CUDA architecture. CUDA has several threads in a single block, whereas each grid contains several such blocks [72, 73]. By assigning image data and by using the thread IDs, the data is processed in a parallel manner.

Recent studies show that researchers prefer hardware accelerators along with software libraries for their visualization experiments. Weinrich et al. [33, 65] used CUDA and OpenGL‡, along with two GPUs, namely, GeForce 8800 GTX and Quadro fx 5600, for their visualization experiments. The

* OpenCL: open computing language.
† AMD: advanced micro devices.
‡ OpenGL: open graphics language.

speed was increased up to 148 times higher than CPU. These were earlier versions of CUDA and OpenGL. At the time of writing this paper, NVIDIA has released a new CUDA version: CUDA 10.0. Recent improvements in CUDA include support for various libraries such as deep learning for medical images [34]. Libraries such as cuDNN, GIE, cuBLAS, cuSPARSE, and NCCL support deep learning in CUDA. The 4D processing of medical images is supported by some of the CUDA libraries (CUFFT and NPP). The present study demonstrated latencies in the visualization pipeline using GPU [35]. In this study, the visualization algorithm is designed using-3Dtexture mapping (3DTM), software-based raycasting (SOFTRC), and hardware-accelerated raycasting (HWRC). They used three GPUs to compare the performance gain. Recently, NVIDIA introduced an RTX* feature in its flagship GPU RTX2080, which is beneficial for direct as well as indirect volume rendering [71].

8.5.2 Advances in TFs

Designing TFs is a complex task. Developers need to consider several parameters (color, opacity, and texture, for instance) while designing an efficient TF. A TF is used to create volumetric data and to extract optical properties (color and opacity, for instance) from volume [60]. The TF is a mandatory part of the visualization pipeline. Medical volume data is a scalar entity, and it is in the 3D spatial domain. 3D image generation involves mapping data from voxel to pixel through a TF. A TF can be categorized into two types: image-centric and data-centric. More generally, the data-centric TF can be designed in four different ways: manual, semiautomatic, automatic, and machine learning–based. In manual TFs, handcrafted parameters are used, so these are not considered for further discussion. In semiautomatic partial user, intervention is required (for view selection), whereas in automatic TF no user intervention is required and the machine learning–based TF learns the data and extracts all the parameters on its own. Researchers have designed several TFs, including image-centric and data-centric TFs which consider all the parameters and design challenges.

8.5.2.1 Image-Centric TFs

In image-centric TFs, the parameters are calculated from the resultant images. These are used to generate a final result from the initial results. Usually, image-centric TFs extract the external shape of objects as a parameter. However, some additional parameters (texture, for instance) are not extracted with precision, which means they miss the minute details. Such minute details are the primary requirement of almost all visualization applications in the medical field, because they provide better insights and are

* RTX: real-time raytracing

useful to identify exact scenarios such as cancerous cells, damages to blood veins and arteries, and minute fractures in bones, for instance. Thus, image-centric TFs are not preferable for medical visualizations [60].

8.5.2.2 Data-Centric TFs

Unlike in image-centric TFs, in data-centric TFs parameters are derived from original data instead of resultant images. To do that, the information from voxels consulting the volume is taken into consideration [60].

Recent research has revealed that volume rendering is a crucial stage in volume visualization. The result of rendering is highly based on the selection of a suitable TF. However, the practical design of the TF is a more complex and time-consuming task. Recently, the research effort has paved the way for semiautomatic and automatic TFs. These TFs are being used to overcome several design challenges in volume visualization. The primary aim of TFs is to identify the objects in underlying data and extract the desired parameters. In addition to this, these parameters are used based on material and projection.

We studied recent advances in TFs that are useful for medical visualization. The following sub-section investigates some of the recent designs of semiautomatic, automatic, and machine learning–based TFs. Earlier research shows that TFs used histogram values of voxels being rendered. Such TFs with scalar values (shown in Equation 8.1) are usually called 1D-TFs.

$$q(d) = C\left(M(d)\right) \tag{8.1}$$

However, by only using histogram values, 1D-TFs can not classify objects properly. Along with histogram values, gradient magnitudes are also required. 2D-TFs based on intensity values and gradient magnitudes (as shown in Equation 8.2) were proposed. They were more effective in detecting multiple materials as well as their boundaries [36].

$$q_{\text{separable}} = (d_1, d_2) = V(M(d_1), d_2) \tag{8.2}$$

Some TFs are considered derivatives of scalar value, along with curvature value [37], feature dimensions [38], and occlusion for ambience [39]. To highlight vital patterns of volume data, a few visibility-based methods were used[40]. Equation 8.3 shows the parameters used in the baseline TF.

$$I = \int_a^b q(s) e^{-\int_a^s K(u)du} ds, \tag{8.3}$$

where I = light intensity between *a* and *b*. The *a* and *b* are traversing points in the volume. $q(s)$= light distribution. *K* is the attenuation of the light. The material properties and light transportation are redefined by functions $q(s)$

and $K(u)$ respectively. The TFs are used to estimate $q(s)$ and $K(u)$. Since baseline TFs are independent of physical properties, baseline TFs are unable to represent objects as ray points and their color values, which represent optical properties of the object. Max et al. [41] proposed Equation 8.4 to include the colors of ray points and their opacity in TF.

$$I = \sum_{i=1}^{n} C_i \alpha_i \prod_{j=1}^{i-1} (1 - \alpha j), \tag{8.4}$$

where C_i is the color of the ray points and α_i is the fraction of light of points in the ray, C_i can be solved recursively in reverse order by Equations 8.5 and 8.6.

$$C'_{i+1} = C'_i + (1 + \alpha'_i) C_i \alpha_i \tag{8.5}$$

$$\alpha'_{i+1} = \alpha'_i + (1 + \alpha'_i) \alpha_i, \tag{8.6}$$

where C_i is accumulated color and α_i is the opacity of the ray points which gives weighted color and opacity distribution. To make TFs design effective and usable across the rendering techniques the semiautomatic, automatic, and machine-learning-based approaches have been proposed by researches in recent studies. We will discuss these studies in the next sub-sections.

8.5.2.2.1 SemiAutomatic TFs

Initially, He et al. [42] proposed an approach to semiautomatic TFs. For the generation of TFs, they used a genetic algorithm. This algorithm could identify the best fit TF, which could be as either user-centric or system-centric for each iteration. The component-based approach proposed by Castro et al. [43] uses a different TF for each of the components such as bones, tissues, and soft flesh. A weighted mixture of this component was used to design a TF. In an attempt to design a new semiautomatic TF, Fang et al. [44] used image enhancement and boundary detection for the transformation of the dataset, which was followed by a linear color operation. This two-step process was useful in designing a semiautomatic algorithm.

Durkin [45] proposed a multi-dimensional volume of attribute values and directional derivatives to construct a histogram. Along with the histogram, they considered the object boundaries in the images. To construct a semiautomatic TF, a model with object boundaries and histogram information has been used to build an opacity function. Further, Prauchner et al. [46] extended their approach and evaluated the various settings. To classify the dataset, a histogram of values or gradients along with 3D coordinations have been used by Roettger et al. [47]. The visibility of hidden patterns was enhanced by optimizing the ranges of absolute values and modulating the opacity using the proposed algorithm inCorrea and Ma [48]. This approach used the histogram for visibility of raypoints. In later work, they added iterative mode views to the visibility data.

The use of clustering techniques to segregate the object features has been implemented to enhance the performance of TFs. Several researchers used clustering techniques to detect the boundaries of objects and materials using similarity amongst the boundary information. Two kinds of clusters, one for identifying boundaries and another for identifying their connectivity in the volumes, were proposed by Šereda et al. [49]. The hierarchy of clusters was used to design a semiautomatic TF. A TF for abdominal visualization with-components generated by a cluster of features was proposed by Maciejewski et al. [50]. They used both value and gradient as features for building a TF. The semiautomatic transfer of properties was studied by some authors instead of the semiautomatic design of TF. Such a transfer of properties is designed in Praßni et al. [51]. This worked as an aid in the rendering of objects in various volumes. The properties were identified during the preprocessing of data.

A hierarchical structure of segments was proposed by Ip et al. [52]. The user could select the appropriate segment of a histogram to generate a TF for each selection. Recently, Liu et al. [53] proposed a method with data containing multiple voxel features. These multiple features can be used to present dynamic projections so that the user can have multiple views of features space that can be used to select aTF.

As thumbnails, semiautomatic TFs require user intervention. Hence, they are not suitable for real-time volume rendering. The volume rendering process gets slower due to user interaction.

8.5.2.2.2 Automatic TF

Semiautomatic TFs require some user interaction, whereas the automatic TF does not require user intervention. An approximate deviation of spatial values of the surface to be rendered was calculated from the mean surface, and it was colored by using the automatic TF in Pfaffelmoser et al. [54]. Wang et al. [55] proposed a simple method to assign color and opacity values to the set of cells. The set of cells isformed by decomposing the feature space extracted from the volume. The cells are separated, and only cells with vital features are kept for further processing. In addition to this, the cells that are derived from noisy data are rejected. Before assigning color and opacity to the cells, a hierarchical structure was created by merging the cells successively. Similarly, 3D field topology and graph representation–based automatic color assignment of TF was proposed in Fujishiro et al. [56]. An importance-based function was used to select the TF automatically as proposed in Wang and Kaufman [57]. In the proposed method, the important color features were used to calculate the importance of objects.

Multi-functional automatic TFs were proposed by Bramon et al. [58]. The information and divergence factors were used to define the object properties while designing the TF. The intensity and gradient magnitude–based automatic method was proposed by Tianjin [60]. Along with value and gradient, they also considered using voxel information and itsspatial arrangement. Finally, they used clustering for generating the TFs.

In summary, automatic TFs are superior to semiautomatic TFs. They fasten the rendering process, and the performance is also better. Most of them use a clustering technique to design TFs.

8.5.2.2.3 Machine Learning–Based TFs

Machine learning helps developers to come up with learned parameters of TFs that are used to project the objects along with their properties. In this type of TFs, the supervised machine learning algorithm [75] uses the training data and the transformation of training data to learn the TF. During this process, the underlying objects are classified, and then TFs are determined. For this purpose, researchers have used various machine learning techniques such as artificial neural networks, hidden Markov models, support vector machines, and clustering techniques.

Various methods based on machine learning have been proposed, and their effects on automatic TFs have been studied by Sundararajan et al. [62]. Their study suggests that the random forest approach is quite appropriate for the design of TFs. An unsupervised learning technique to reduce the dimensionality of volumes and TF space is proposed in De Moura Pinto and Freitas [59]. The proposed technique also used unsupervised machine learning to assign object properties such as color and opacity on the reduced dimensions. The artificial neural network with back propagation has been used by Wang et al. to find the similarities in the input volumes [64]. The TF is generated by classifying the information from a set of parameters. Selver et al. [61] proposed a method in which both high- and low-frequency structures were represented in different quadrants of transformation for enhancement of volume data. They used brushlet expansion for volume reconstruction with tiling of selected quadrants.

In recently studied techniques, it has been observed that Gaussian-based naive Bayesian classifiers have limited offerings for the complex dataset and that they cannot handle the effect of outliers, although they are quite fast for classification. Significant accuracy is achieved by using k-nearest neighbors (k-NN) classifiers. However, they are quite expensive in terms of computational cost, since they cannot obtain high-dimensional data from the input training data [62].

Similarly, in the case of the support vector machine (SVM), the training time is high, but it is faster than k-NN; for the classification provided the data should be normalized. Single-layer perceptrons also provedto be good, but they are not comparable with other classifiers. Machine learning–based TFs have benefited from simple Bayesian networks, but random forests have been significant in all kinds of challenges related to TFs.

In summary, there are various other categories of TFs based on the design aspect. Many of them require acomplicated method to design and estimate the parameters of the TF. Such methods cannot be considered as an advancement. More information on these methods is available in Ljung et al. [63].

8.5.3 Generative Adversarial Networks(GANs)

GANs are proving to be milestones, particularly in image processing domains. A GAN generator network generates images based on supervised training data, whereas an adversarial network assesses generated images. The learning function adjusts the generator network's parameters based on the advice received from the adversarial network. Figure 8.9 shows the application of GAN to generate visualization data.

A view variant TF space is learned using a generative network, which quantifies the desired changes to be made by analyzing the output image in the training data [68]. The generated space is used directly for rendering, which enables the user to explore the entire space of generated images. Since the model is independent of the rendering process, the algorithm also demonstrates resultant images generated by global illumination lighting on various datasets. A new interactive visualization tool (GAN Lab) can be used to experiment with GAN using selective and popular deep learning models. It allows users to train various generative models and also to visualize intermediate results during the training process [69].

Recently, a GAN model with various stages was proposed to generate 3D volumes for a variety of objects [70]. The generator network captures the object structure and produces 3D objects with high quality. It also creates a mapping between probabilistic space and3D space for objects. This can be done without using a reference image or external models (CAD, for instance). The discriminator network in the model creates a 3D descriptor for the shape, which is learned without any supervision; such a model can be used to recognize 3D objects in the generated volumes.

In summary, looking at the advancements shown by GAN models, it is esteemed that they can play a vital role in medical imaging and visualization. More and more research is expected tomake use of GANs to generate, render, and recognize 3D data.

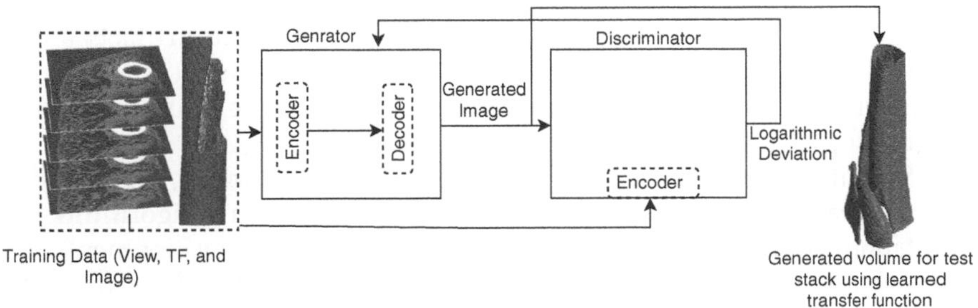

FIGURE 8.9
How GAN works with labeled images to learn the visualization transformation.

8.6 Tools and Libraries for Volume Visualization

Several software tools and graphics libraries are available for volume visualization. Most of them are open source. DICOM viewers are commonly used by the radiographer to visualize the volume data acquired by 3D scanning devices. They have the provisions of 3D model generation from that data. Other than this, 3D Slicer and 3D-DOCTOR are the tools generally used by researchers to reconstruct 3D models. The models generated by these tools can be stored in the STL file for future use. A blender tool is used in Johns [30] to develop an augmented reality–based driving simulator. In addition to this, CUDA, OpenCL, OpenGL, VolPack, and Visualization Toolkit (VTK) are the libraries available to do effective volume visualization.

8.7 Conclusion and Future Directions

In this paper, we presented the systematic review on several volume rendering techniques including both direct and indirect volume rendering. The typical steps involved in the volume visualization process are discussed in detail. Recent advances in TF, which include automatic and semiautomatic designs, are discussed in detail. Along with this, we discussed machine learning–based TFs and their performance. Hardware-based advancements such as CUDA programming using various GPUs and their effect on visualization processing are also specified. In addition to this, a GAN-based architecture for realistic visualization is discussed in brief.

In the data classification (segmentation) step, most of the previous attempts used thresholding-based approach. Thresholding is a traditional segmentation technique that neither analyses the image contents nor assigns unique labels (required in case there are multiple objects present in the volume). In the future, more research attempts should be made to devise image content analysis–based preprocessing and segmentation techniques. The preprocessing technique should be able to remove unwanted artifacts and to enhance the required region of interest by analyzing image contents. The segmentation technique should not only extract the desired portion, but also assign unique labels to each object in the volume.

In the future, researchers can think about using automatic (deep and/or machine learning–based) TFs to develop volume rendering algorithms. The adaptation of automatic TFs will definitely reduce the development time; moreover, researchers can focus on the development of the desired system. Other than this, more research must be conducted to design hybrid volume visualization techniques which initially render volume by surface rendering approach, and after detecting a collision, the technique should generate internal details on the fly through DVR to perform erosion logic.

References

1. P. Reddy, M. Balaram, C. Bones, and Y. B. Reddy, *Visualization in Scientific Computing*, Grambling State University, Grambling, LA, 1996.
2. A. Kaufman, "Volume visualization," *The Visual Computer*, vol. 6, no. 1, pp. 1–1, 1990.
3. B. H. McCormick, T. A. DeFanti, and M. D. Brown (eds), "Visualization in scientific computing," special issue of *Computer Graphics*, ACM, vol. 21, no. 6, 1987.
4. T. A. Defanti and M. D. Brown, "Visualization in scientific computing," *Advances in Computers*, vol. 33, pp. 247–307, 1991.
5. D. D. Ruikar, R. S. Hegadi, and K. C. Santosh, "A systematic review on orthopedic simulators for psycho-motor skill and surgical procedure training," *Journal of Medical Systems*, vol. 42, no. 9, p. 168, 2018.
6. A. Kaufman, "Volume visualization: principles and advances," Course notes, 24, 1997.
7. T. T. Elvins, "A survey of algorithms for volume visualization," *ACM Siggraph Computer Graphics*, vol. 26, no. 3, pp. 194–201, 1992.
8. Q. Zhang, R. Eagleson, and T. M. Peters, "Volume visualization: a technical overview with a focus on medical applications," *Journal of Digital Imaging*, vol. 24, no. 4, pp. 640–664, 2011.
9. J. Zhou and K. D. Tonnies, "State of the art for volume rendering," *Simulation*, pp. 1–29, 2003.
10. H. Pfister and C. M. Wittenbrink, Volume visualization and volume rendering techniques, Tutorial presented at Eurographics, 2000.
11. C. D. Hansen, and C. R. Johnson, *Visualization Handbook*, Elsevier, 2011.
12. W. E. Lorensen and H. E. Cline, "Marching cubes: a high-resolution 3D surface construction algorithm," *Computer Graphics*, vol. 21, no. 4, pp. 163–169, 1987.
13. M. Levoy, "Volume rendering by adaptive refinement," *The Visual Computer*, vol. 6, no. 1, pp. 2–7, 1990.
14. M. Levoy, "Efficient ray tracing of volume data," *ACM Transactions on Graphics (TOG)*, vol. 9, no. 3, pp. 245–261, 1990.
15. S. Grimm, S. Bruckner, A. Kanitsar, and E. Gröller, "A refined data addressing and processing scheme to accelerate volume raycasting," *Computers & Graphics*, vol. 28, no. 5, pp. 719–729, 2004.
16. L. Westover, "Interactive volume rendering," in *Proceedings of the 1989 Chapel Hill Workshop on Volume Visualization*, ACM, 1989, pp. 9–16.
17. L. Westover, "Footprint evaluation for volume rendering," *ACM Siggraph Computer Graphics*, vol. 24, no. 4, pp. 367–376, 1990.
18. K. Subr, P. Diaz-Gutierrez, R. Pajarola, and M. Gopi, Order independent, attenuation-leakage free splatting using freevoxels, Tech. Rep. IFI-2007.01, Department of Informatics, University of Zürich, 2007.
19. M. Zwicker, H. Pfister, J. Van Baar, and M. Gross, "EWA splatting," *IEEE Transactions on Visualization and Computer Graphics*, vol. 8, no. 3, pp. 223–238, 2002.
20. D. Xue and R. Crawfis, "Efficient splatting using modern graphics hardware," *Journal of Graphics Tools*, vol. 8, no. 3, pp. 1–21, 2003.
21. J. P. Schulze, M. Kraus, U. Lang, and T. Ertl, "Integrating pre-integration into the shear-warp algorithm," in *Proceedings of the* 2003 Eurographics/IEEE TVCG Workshop on Volume Graphics, ACM, 2003, pp. 109–118.

22. H. Kye and K. Oh, "High-quality shear-warp volume rendering using efficient supersampling and pre-integration technique," in *Advances in Artificial Reality and Tele-Existence*, 2006, pp. 624–632, Springer, Berlin, Heidelberg, Germany.
23. P. S. Calhoun, B. S. Kuszyk, D. G. Heath, J. C. Carley, and E. K. Fishman, "Three-dimensional volume rendering of spiral CT data: theory and method," *Radiographics*, vol. 19, no. 3, pp. 745–764, 1999.
24. L. Mroz, H. Hauser, and E. Gröller, "Interactive high-quality maximum intensity projection," in *Computer Graphics Forum* (Vol. 19, No. 3), 2000, pp. 341–350, Blackwell Publishers Ltd, Oxford, UK and Boston, MA.
25. T. J. Cullip and U. Neumann, *Accelerating Volume Reconstruction with 3D Texture Hardware*, 1993, University of North Carolina at Chapel Hill. Department of Computer Science.
26. B. Cabral, N. Cam, and J. Foran, "Accelerated volume rendering and tomographic reconstruction using texture mapping hardware," in *Proceedings of the 1994 Symposium on Volume Visualization*, ACM, 1994, pp. 91–98.
27. A. Van Gelder and K. Kim, "Direct volume rendering with shading via three-dimensional textures," in *Proceedings of the 1996 Symposiumon Volume Visualization*, IEEE, 1996, pp. 23–30.
28. C. Rezk-Salama, K. Engel, M. Bauer, G. Greiner, and T. Ertl, "Interactive volume on standard PC graphics hardware using multi-textures and multi-stage rasterization," in *Proceedings of the* ACM SIGGRAPH/EUROGRAPHICS Workshop on Graphics Hardware, ACM, 2000, pp. 109–118.
29. P. Abellán and D. Tost, "Multimodal volume rendering with 3D textures," *Computers & Graphics*, vol. 32, no. 4, pp. 412–419, 2008.
30. B. D. Johns, The creation and validation of an augmented reality orthopaedic drilling simulator for surgical training. Thesis, University of Iowa, 2014.
31. W. E. Lorensen and H. E. Cline, "Marching cubes: a high-resolution 3D surface construction algorithm," in *ACM Siggraph Computer Graphics* (Vol. 21, No. 4), ACM, 1987, pp. 163–169.
32. G. M. Treece, R. W. Prager, and A. H. Gee, "Regularised marching tetrahedra: improved iso-surface extraction," *Computers & Graphics*, vol. 23, no. 4, pp. 583–598, 1999.
33. A. Weinlich, B. Keck, H. Scherl, M. Kowarschik, and J. Hornegger, "Comparison of high-speed ray casting on GPU using CUDA and OpenGL," in *Proceedings of the First International Workshop on New Frontiers in High-performance and Hardware-aware Computing* (Vol. 1), Proceedings of HipHaC'08, 2008, pp. 25–30.
34. NVIDIA. Deep learning software. Available at: https://developer.nvidia.com/deep-learning-software. Accessed October 29, 2018.
35. Q. Zhang, R. Eagleson, T. M. Peters, "Dynamic real-time 4D cardiac MDCT image display using GPU-accelerated volume rendering," *Computerized Medical Imaging and Graphics*, vol. 33, no. 6, pp. 461–476, 2009.
36. G. Kindlmann and J. W. Durkin, "Semi-automatic generation of transfer functions for direct volume rendering," in 1998 IEEE *Symposium on Volume Visualization*, IEEE, 1998, pp. 79–86.
37. G. Kindlmann, R. Whitaker, T. Tasdizen, and T. Möller, "Curvature-based transfer functions for direct volume rendering: methods and applications," in *Proceedings of the 14th* IEEE Visualization (*VIS '03*), IEEE, 2003, pp. 513–520.

38. C. D. Correa and K.-L. Ma, "Size-based transfer functions: a new volume exploration technique," *IEEE Transactions on Visualization and Computer Graphics*, vol. 14, no. 6, pp. 1380–1387, 2008.
39. C. D. Correa and K.-L. Ma, "The occlusion spectrum for volume classification and visualization," *IEEE Transactions on Visualization and Computer Graphics*, vol. 15, no. 6, pp. 1465–1472, 2009.
40. C. D. Correa and K.-L. Ma, "Visibility histograms and visibility-driven transfer functions," *IEEE Transactions on Visualization and Computer Graphics*, vol. 17, no. 2, pp. 192–204, 2011.
41. N. Max, "Optical models for direct volume rendering," *IEEE Transactions on Visualization & Computer Graphics*, vol. 1, no. 2, pp. 99–108, 1995.
42. T. He, L. Hong, A. Kaufman, and H. Pfister, "Generation of transfer functions with stochastic search techniques," in *Visualization'96. Proceedings*, IEEE, 1996, pp. 227–234.
43. S. Castro, A. König, H. Löffelmann, and E. Gröller, Transfer function specification for the visualization of medical data, Vienne University of Technology, 1998.
44. S. Fang, T. Biddlecome, and M. Tuceryan, "Image-based transfer function design for data exploration in volume visualization," in *IEEE Visualization*, 1998, pp. 319–326.
45. G. Kindlmann and J. W. Durkin, "Semi-automatic generation of transfer functions for direct volume rendering," in *1998 IEEE Symposium on Volume Visualization*, IEEE, 1998, pp. 79–86.
46. J. L. Prauchner, C. M. Freitas, and Comba, J. L. D., "Two-level interaction approach for transfer function specification," in *Computer Graphics and Image Processing*, 2005, SIBGRAPI, 18th Brazilian Symposium on, IEEE, 2005, pp. 265–272.
47. S. Roettger, M. Bauer, and M. Stamminger, "Spatialized transfer functions," in *Euro Vis*, 2005, pp. 271–278.
48. C. D. Correa and K. L. Ma, "Visibility-driven transfer functions," in *Visualization Symposium*, 2009. *Pacific Vis'09*. IEEE Pacific, IEEE, 2009, pp. 177–184.
49. P. Sereda, A. Vilanova, and F. A. Gerritsen, "Automating transfer function design for volume rendering using hierarchical clustering of material boundaries," in *Euro Vis*, 2006, pp. 243–250.
50. R. Maciejewski, I. Woo, W. Chen, and D. Ebert, "Structuring feature space: a non-parametric method for volumetric transfer function generation," *IEEE Transactions on Visualization and Computer Graphics*, vol. 15, no. 6, pp. 1473–1480, 2009.
51. J. S. Praßni, T. Ropinski, J. Mensmann, and K. Hinrichs, "Shape-based transfer functions for volume visualization," in *Visualization Symposium (Pacific Vis)*, 2010 IEEE Pacific, IEEE, 2010, pp. 9–16.
52. C. Y. Ip, A. Varshney, and J. JaJa, "Hierarchical exploration of volumes using multilevel segmentation of the intensity-gradient histograms," *IEEE Transactions on Visualization & Computer Graphics*, vol. 18, no. 12, pp. 2355–2363, 2012.
53. S. Liu, B. Wang, J. J. Thiagarajan, P. T. Bremer, and V. Pascucci, "Multivariate volume visualization through dynamic projections," in *2014 IEEE 4th Symposium on Large Data Analysis and Visualization (LDAV)*, IEEE, 2014, pp. 35–42.
54. T. Pfaffelmoser, M. Reitinger, and R. Westermann, "Visualizing the positional and geometrical variability of isosurfaces in uncertain scalar fields," in Computer Graphics *Forum* (Vol. 30, No. 3), Blackwell Publishing Ltd, Oxford, UK, 2011, pp. 951–960.

55. Y. Wang, J. Zhang, D. J. Lehmann, H. Theisel, and X. Chi, "Automating transfer function design with valley cell-based clustering of 2D density plots," in Computer Graphics *Forum* (Vol. 31, No. 3 pt 4), Blackwell Publishing Ltd, Oxford, UK, 2012, pp. 1295–1304.
56. I. Fujishiro, Y. Takeshima, T. Azuma, and S. Takahashi, "Volume data mining using 3D field topology analysis," *IEEE Computer Graphics and Applications*, vol. 20, no. 5, pp. 46–51, 2000.
57. L. Wang and A. Kaufman, "Importance driven automatic color design for direct volume rendering," in Computer Graphics *Forum* (Vol. 31, No. 3 pt 4), Blackwell Publishing Ltd, Oxford, UK, 2012, pp. 1305–1314.
58. R. Bramon, M. Ruiz, A. Bardera, I. Boada, M. Feixas, and M. Sbert, "Information theory-based automatic multimodal transfer function design," *IEEE Journal of Biomedical and Health Informatics*, vol. 17, no. 4, pp. 870–880, 2013.
59. F. De Moura Pinto and C. M. Freitas, "Design of multi-dimensional transfer functions using dimensional reduction," in *Proceedings of the 9th Joint Eurographics/IEEE VGTC Conference on Visualization*, Eurographics Association, 2007, pp. 131–138.
60. T. Zhang, Z. Yi, J. Zheng, D. C. Liu, W. M. Pang, Q. Wang, and J. Qin, "A clustering-based automatic transfer function design for volume visualization," *Mathematical Problems in Engineering*, 2016.
61. M. A. Selver, "Exploring brushlet based 3D textures in transfer function specification for direct volume rendering of abdominal organs," *IEEE Transactions on Visualization and Computer Graphics*, vol. 21, no. 2, pp. 174–187, 2015.
62. K. P. Soundararajan and T. Schultz, "Learning probabilistic transfer functions: a comparative study of classifiers," in *Computer Graphics Forum* (Vol. 34, No. 3), 2015, pp. 111–120.
63. P. Ljung, J. Krüger, E. Groller, M. Hadwiger, C. D. Hansen, and A. Ynnerman, "State of the art in transfer functions for direct volume rendering," in *Computer Graphics Forum* (Vol. 35, No. 3), 2016, pp. 669–691.
64. L. Wang, X. Chen, S. Li, and X. Cai, "General adaptive transfer functions design for volume rendering by using neural networks," in *International Conference on Neural Information Processing*, Springer, Berlin, Heidelberg, 2006, pp. 661–670.
65. A. Weinlich, B. Keck, H. Scherl, M. Kowarschik, and J. Hornegger, "Comparison of high-speed ray casting on GPU using CUDA and OpenGL," in *Proceedings of the First International Workshop on New Frontiers in High-performance and Hardware-aware Computing* (Vol. 1), Proceedings of HipHaC'08, 2008, pp. 25–30.
66. Q. Zhang, R. Eagleson, and T. M. Peters, "Dynamic real-time 4D cardiac MDCT image display using GPU-accelerated volume rendering," *Computerized Medical Imaging and Graphics*, vol. 33, no. 6, pp. 461–476, 2009.
67. M. Berger, J. Li, and J. A. Levine, "A generative model for volume rendering," *IEEE Transactions on Visualization & Computer Graphics*, no. 1, pp. 1–1.
68. M. Kahng, N. Thorat, D. H. P. Chau, F. B. Viégas, and M. Wattenberg, "GAN lab: understanding complex deep generative models using interactive visual experimentation," *IEEE Transactions on Visualization and Computer Graphics*, 2018.
69. J. Wu, C. Zhang, T. Xue, B. Freeman, and J. Tenenbaum, "Learning a probabilistic latent space of object shapes via 3D generative-adversarial modeling," In *Advances in Neural Information Processing Systems*, 2016, pp. 82–90.
70. NVIDIA. GE Forxe RTX. Available at: https://www.nvidia.com/en-us/geforce/20-series/rtx/.

71. Nvidia CUDA, *Nvidia Cuda c Programming Guide*, Nvidia Corporation, 120(18), 8, 2011.
72. T. Kalaiselvi, P. Sriramakrishnan, and K. Somasundaram, "Survey of using GPU CUDA programming model in medical image analysis," *Informatics in Medicine Unlocked*, vol. 9, pp. 133–144, 2017.
73. D. D. Ruikar, R. S. Hegadi, and K. C. Santosh, Contrast stretching-based unwanted artifacts removal from CT images in recent trends in image processing and pattern recognition (accepted), Springer, 2019.
74. S. Vajda and K. C. Santosh, "A fast k-nearest neighbor classifier using unsupervised clustering," in *International Conference on Recent Trends in Image Processing and Pattern Recognition*, Springer, Singapore, 2016, pp. 185–193.

9

A Review on the Evolution of Comprehensive Information for Digital Sliding of Pathology and Medical Image Segmentation

M. Ravi and Ravindra S. Hegadi

CONTENTS

9.1 Introduction

Characterized as the act of diagnosing disease, pathology is a focal subspecialty in diagnosis. Anatomic pathology, a part of pathology, utilizes microscopy to picture and analyze the malady procedure in tissue. Notwithstanding the fruitful and boundless utilization of imaging information and automated procedures in other medicinal services fields, such as radiology, the fundamental procedures by which analyses are made in anatomic pathology remain for the most part without programmed intervention. In any case, late advances in computerized pathology imaging, particularly in the field of entire slide imaging, have started the change to computerized pathology rehearse. As digitalization gadgets turn out to be more reasonable, their utilization in pathology research facilities is expanding [1,2]. The College of American Pathologists built a revolutionary program keeping in Intellect to help speed up a selection of involuntary advances in pathological research centers [3]. Furthermore, the association of pathologists has logically changed the examination to embrace computerized pictures, leaving a few inquiries subordinate upon coordinate examination of glass slides utilizing a magnifying lens [4]. As this innovation is actualized, pathology imaging information is developing at an exponential rate and is anticipated to far surpass the tremendous amounts of imaging information.

9.1.1 Hurdles Encountered in Digitization of Pathological Information

A complete utilization of computerized pathology imaging information has the guarantee to enhance understanding, yet critical technological challenges remain that will push the breaking points of media and correspondences.

Picture dimensions: a run of the whole slide pathology picture contains 20 billion pixels. A 24-b shading un-packed depiction of this normal picture contains 56 GB. Pressure with JPEG commonly decreases this size to a few GB or several MB. A multidimensional pathological picture has images stacked together to provide z-depth, which in turn provides more insight into the picture. A solitary checking framework can quality rate many pictures every day, which makes a noteworthy problem for capacity and investigation.

Low latency: it gets to pathological images that are normally kept up on a brought together server and seen remotely by professionals over the system. Low-idleness serving is required to keep up a liquid survey encounter that allows the panning and zooming activities required for indicative methods. Most conclusions are rendered offline through the examination of a biopsy, yet once in a while secluded considering is performed continuously amid surgical intercession.

Unique compression necessities: the substance of pathology pictures presents extraordinary difficulties for picture pressure. Pressure relics go up against significance when the compacted content is being utilized to make a conclusion. Forceful compression is a prerequisite for liquid review; however, the impacts of existing strategies created for photography applications are as yet not well characterized. Furthermore, pressure plans for multilayer pictures that adventure z-repetition are not right now spoke of in business frameworks.

These difficulties are exacerbated by the development of an extra layer of pathology media: quantitative information created by computational picture investigation. This inferred information speaks to the substance of a virtual pathology slide at different resolutions, from singular portrayals of every one of a large number of cells to local/textural depictions of entire tissues. The execution, administration, and de-uniform of this extra layer of substance works in advance for symptomatic applications. The advancements in medical imaging and storage can lead to milestones in pathological image analysis. Picture investigation will assume a basic part in the distinguishing proof of novel helpful targets and production of new ailment order frameworks that enhance forecast of treatment reaction. Similarly, entire genome sequencing gives a window into a patient's hereditary inclination to illness and capacity of treatment reaction. The computation of highlights speaking to a huge number of carefully gathered tissue pictures will light up the interaction amongst particles and organic substances. Picture investigation information will give a rich storehouse where designs that anticipate survival and treatment reaction can be derived while additionally clarifying malady components and revolutionizing the care of patients with tumors and different infections.

This chapter is divided into six sections. Section 9.1 is the introduction, Section 9.2 gives the foundation on the act of pathology including a concise history of microscopy and its effect on setting up demonstrative methods. Section 9.3 talks about digitization in pathology imaging and gives a study of the cutting edge in advanced modalities. In Section 9.4, pathology picture examination and momentum inclines in this re-seek territory are talked about. Section 9.5 portrays the segment related to advanced pathology practice and picture using image processing for pathological medical imaging. Section 9.6 is the conclusion.

9.2 Pathology Origins towards Whole Slide Imaging (WSI)

Anatomic pathology is a subdivision within the field of pathology that has generally utilized magnifying instruments to analyze the cell and sub-cellular morphology of examples, either as strong tissues or liquids. At the point when a sample is taken from a patient, it is sent to the pathology research facility for examination and conclusion. Research facility staff first outwardly analyzes the specimen without the guide of a magnifying instrument, and this is called net examination, where prepared workforce selects a tiny tissue for examination. The digital slides of each protein tissue are examined under the instrument. After initial examination is finished, the tissue is additionally prepared to supplant water with natural solvents and is then implanted in a piece of paraffin wax. Thin cuts of the tissue are mounted on glass slides, and these slides are taken through a progression of mix stains, every one of which has a penchant for select cell components. In the wake of stain, a drop of optical review mounting paste is put over the tissue taken after by a glass or plastic cover slip, both of which consider a clear investigation of the recolor tissue area while shielding it from harm. Stains do not just enable the pathologist to see the tissue under the magnifying instrument, they also upgrade the location of particular changes in the tissue that happen because of sickness. The most utilized stain is a mix of hematoxylin and eosin (H&E) that features miniaturized scale anatomic structures. Hematoxylin is dull purple and has a liking for the nucleic acids exhibited in cell cores, while eosin is red and has a liking for protein introduced in the cytoplasm and extracellular spaces of the tissue (see Figure 9.1).

After slide readiness is finished, they are conveyed with proper clinical data in regards to the patient to the pathologist for infinitesimal examination. By and large, the pathologist starts by looking at the tissue segment on a glass slide at low amplification to get an expansive view, and this is trailed by examination of chosen territories under high amplification for better representation of demonstrative and additionally difficult regions. The pathologist records his or her discoveries as per the momentum standard of watch over the ailment and tissue write in a pathology report. Such discoveries may incorporate assurance of the degree of a tumor, the status of the edges around a tumor (i.e. to decide if a specialist excised the whole tumor), the tally of cells experiencing division (mitosis) per high-amplification field, and the nearness or nonattendance of aggravation, rare specimen, vascular tumor intrusion, and so forth. On the off chance that the determination can't be made with H&E recolor alone, extra stains might be utilized to help settle vulnerabilities. Several different stains are routinely accessible in most pathology research facilities, and pathologists will choose these in light of the presumed conclusion. Immunohistochemical (IHC) stains utilize a counteracting agent that has been created for a particular target, called an antigen, which is labeled with a colorimetric flag (see Figure 9.1). These stains are connected to tissues and inspected for the nearness and area of specific antigenic focuses on that can't generally be seen utilizing a magnifying instrument and routine H&E stains. IHC stains are especially valuable for the discovery of rare living specimen and for strange antigenic changes in cells identified with the tumor.

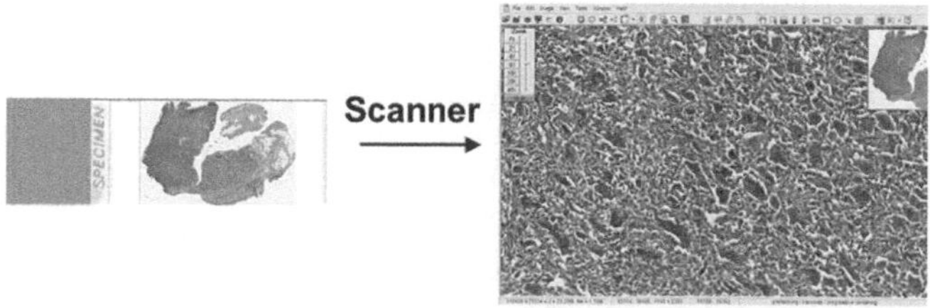

FIGURE 9.1
WSI catches the substance of a whole glass slide at high amplified view.

The examination of fine biologic detail utilizing a lit magnifying instrument started around 500 years back. In 1595, the Dutch display creator Hans Jansen and his child Zacharias built the primary compound magnifying lens comprising of two focal points inside an empty tube, the separation between which could be shifted to accomplish distinctive levels of amplification [5]. Since their work was at first unpublished, great part of the credit went to Robert Hooke, a member of the Royal Society in London who dispersed the primary work on microscopy, *Micrographia,* in 1665 [6]. Here Hooke portrayed his refined compound magnifying instrument, having three optical focal points, a phase, and a light source. Antonie van Leeuwenhoek, a Dutch researcher, made important discoveries in the development of focal points that enhanced amplification, allowing the representation of microscopic organisms, protozoa, and spermatozoa, for which he wound up known as the Father of Microbiology [7, 8].

In the late 1800s and mid-1900s, numerous vital minuscule revelations were made through the utilization of stains and manufactured colors. Joseph von Gerlach depicted the differential recolor of the core and cytoplasm utilizing carmine color in 1858 [7]. This was trailed by the utilization of silver staining by Camillo Golgi in 1873 [7]; the essential fluorescent shading, fluorescein by Adolf von Bayer in 1871 [7, 9]; and the usage of the H&E recolor by Paul Mayer in 1896 [7]. Vital advances in light microscopy were likewise made in the late 1800s, including the elaboration of the possible hypothesis in 1873 by Ernst Abbe, working as a team with Carl Zeiss and Otto Schott [7, 10]. This hypothesis expressed that the littlest resolvable separation between two focuses utilizing an ordinary small-scale extension may never be not as much as a large portion of the imaging light wavelength, and the use of this guideline prompted upgrades in magnifying lens development [7]. Oskar Heimstadt built up the principal pragmatic fluorescence magnifying instrument in 1911 after August Kohler concocted the primary bright magnifying instrument in 1904 [7, 11]. Consequent to this, the advancement of various subordinate methods has empowered the investigation of biologic frameworks including: immune fluorescence for the counteracting agent named recognition of particular antigens [11–14], immune peroxidase methods for distinguishing antigens in standard light tiny segments [15–18], and utilization of green fluorescent protein for the examination of quality articulation [7, 19, 20].

9.3 Digitalization of Pathology Imaging

Use of minuscule information for the better gynecological screening was one of the initial moves as far as the digital pathology was concerned in the late 1950s. The CYDAC Image Cytometric Microscope System, constructed halfway with respect to the Nipkow Disk, is a striking case of early endeavors to digitize pathology hone [21, 22]. Beginning endeavors at advanced pathology were just ready to catch parts of tissue areas for exam-area since amplification blocked the incorporation of the whole tissue segment into a solitary computerized photo. In the end, instruments were created that could catch a whole slide into a solitary computerized picture, called a WSI [2, 21, 29]. At the beginning, ways to deal with WSI were performed by getting different covering advanced photomicrographs gained through the span of numerous hours utilizing a mechanical magnifying instrument and afterward sewing together the pictures at their limits. A virtual magnifying instrument was created that enabled clients to explore WSI at different levels of amplification. Further, WSI allows point magnification that helps to focus on the region of interest rather than looking at the entire slide image [30, 31]. An early WSI instrument was produced by John Gilbertson and Art Wetzel, who were then at Interscope Advancements [32]. From that point forward, countless have turned out to be financially accessible [33], and the rundown of such frameworks is developing. With the accessibility of graphical UIs that permit the execution of a virtual microscope [31] and increments in securing speeds, these gadgets are presently being utilized in research and instruction as well as in day by day pathology rehearse.

Telepathology is the utilization of advanced pathology pictures to perform analysis without glass slides and a magnifying lens, and the coming of WSI innovation has caused a sharp increment in the reception of telepathology for clinical care. The first articles to utilize the term telepathology were published in 1986, before the discovery of WSI [2, 28]. The utilization of WSI frameworks for telepathology has brought about various investigations depicting its adequacy for care [34, 35], and a white paper on telepathology approval is normal in the last 50% of 2012 Digital Pathology Working Group. Norway is one of the countries where frozen and remote telepathology has been progressing effectively [23–27]. An expansion in informatics preparing in pathology residency projects will probably prompt further increments in the use of this innovation [36–38], since pathologists can contribute to enhancement of the information base by uploading clinical information, and it was anticipated that this would be at some point in 2013 [39, 40] (Table 9.1).

TABLE 9.1

Achievements Timeline in Microscopy and Digital Pathology

Sl. No	Year	Achievements in Pathology Digitization
1	1590	Discovery of the principle of the compound microscope by Zacharias Janssen, Dutch spectacle-maker.
2	1665	Robert Hooke publishes *Micrographia.*
3	1911	The first fluorescence microscope is developed by Oskar Heimstadt.
4	1987	Concept of telepathology.
5	1990	An achievement used in microscopic pathology with extensive digitization by CCD and CMOS sensors
6	2000 onwards	WSI is extensively used in digital pathology.

A study of the present market discovers in excess of ten organizations as of now offering WSI arrangements. Contemporary frameworks are offered in an assortment of arrangements, from shut high-throughput line filtering gadgets to cheap scale down to single slide frameworks for work area utilization. The presence of record terms identified with computerized pathology in logical distributions shows the development in the gadgets advertised (see Figure 9.2).

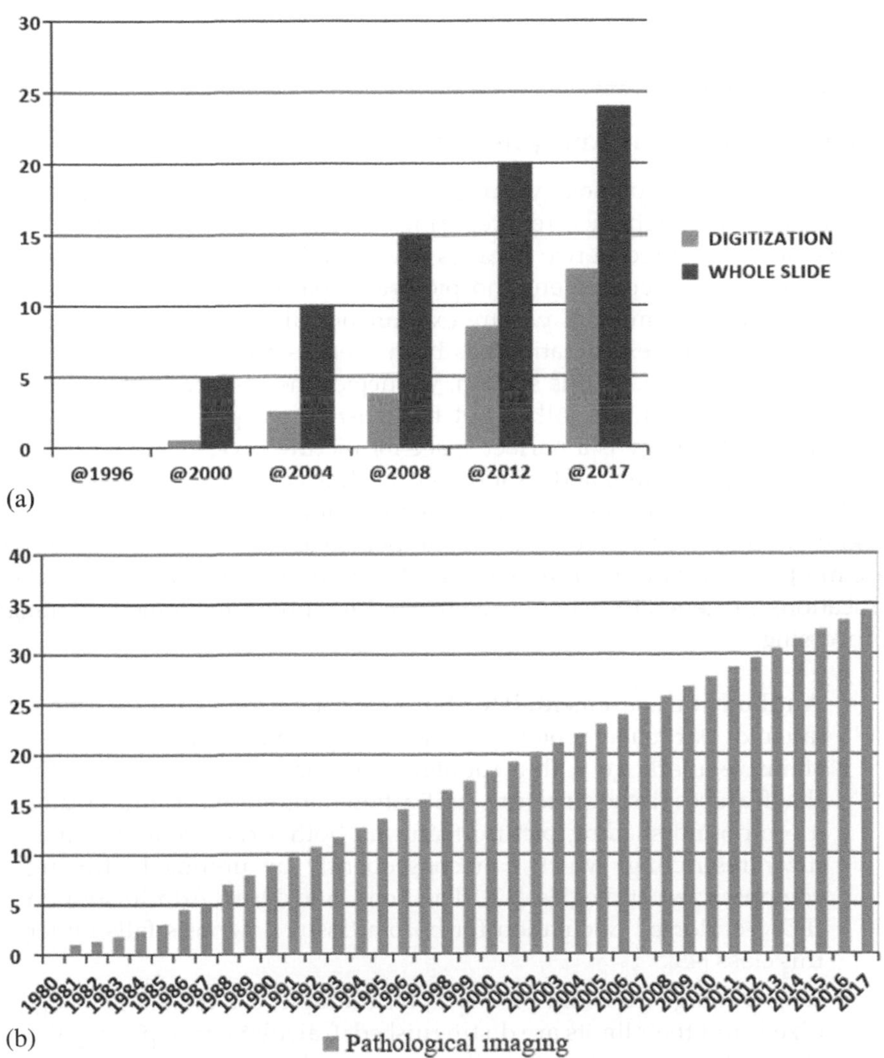

FIGURE 9.2
Medicine record of pathology digitization and picture investigation.

A few slides examining gadgets were presented at the main European Scanner Contest held in May 2010 in Berlin, Germany, to assess velocity, center, and picture quality [41]. Novel concentration components have risen in the most recent age of scanners, with a few gadgets offering dynamic concentration advancements that utilize double sensors or loop-mounted sensors to quicken filtering and enhance center quality. Tape-style slide holders that handle slides in a roundabout way are additionally supplanting the perplexing mechanical technology used to straightforwardly control singular slides, as the last are inclined to misusing of slides and breakdown.

9.4 Computational Analysis of Pathological Imaging

The digitization of pathology imaging opened the doors for quantitative examination through picture preparing. Upgrades in digitization in the course of the most recent two decades have been joined by similar advances in both registering equipment and picture preparation strategy. The result has been a development of picture examination in the pathology imaging area where picture examination has been used as a piece of both research and clinical settings. In this section, we depict the essentials of pathology picture examination and talk about the most recent patterns and current difficulties. Pathology is a perfect space for picture investigation. The ability to section, measure, and group pictures has coordinate applications in routine pathology errands, including quantification of neutralizer reclosing, acknowledgment and order of cells, and portrayal of minute structures that are multicellular or territorial in nature. The accompanying application classifications are generally experienced in a writing program for medical image processing.

- **Antibody measurement.** IHC stains are connected to feature articulation of particular proteins or their transformed structures. These stains regularly go with a counter stain, for example, hematoxylin that features tissue structure. Utilization of numerous stains creates a compound shading picture featuring both structure and immunizer restricting, which can consequently be unmixed utilizing computerized strategies [42]. The Food and Drug Administration (FDA)-endorsed calculation for bosom tissue diagnosis falls under this class [43].
- **Object image division.** Elements, for example, cell cores, are recognized, and their limits are distinguished. Calculations to portray the shape, shading, and surface of a fragmented protest frequently take after division and are a case of highlight extraction.

- **A region of image division.** Regularly, the substances to be segmented are made out of accumulations of straightforward questions and structures and are characterized by a complex or textural appearance. Illustrations incorporate recognizing the limits of veins, injuries, and irritation.
- **Feature extraction.** Highlight extraction is the way toward figuring useful depictions of items or locales and frequently goes before order or division assignments. Feature extraction process represents the object characteristics in a vectorized form so that object can be categorized to particular class robustly using some machine learning methods.
- **Classification.** Sectioned articles, areas, or entire slides can be arranged into important gatherings in light of removed highlights. Based on characteristics present in the slides they can be classified to its appropriate class by means of training the machine learning (ML) models.
- **Registration.** Picture enlistment is the way toward mapping at least two pictures into a similar co-ordinate outline. Enlistment can be utilized to make 3D recreations of tissue from an arrangement of tissue segments or to outline diversely recolored areas to each other to incorporate immune response nearness

A concise foundation of research in pathology picture examination is exhibited beneath. A more inside and out specialized survey is accessible in Gurcan et al. [44].

i) **Image Segmentation:** The divisions of cell nuclei, cell layers, or subcellular segments are the most essential issues in pathology picture examination. Staining can be connected to uncover these articles as uniform locales of particular shading. Shading division is one of the major issues in picture investigation; however, segmentation of pathology pictures keeps being tested because of regular natural and process-instigated varieties crosswise over pictures. The techniques usually utilized incorporate
   - Thresholding
   - Active contours
   - Bayesian methodology
   - Region growing developing
   - Image clustering.

   A typical subject in pathology picture examination is the division of cell cores. A Bayesian atomic segmentation that incorporates shading, surface, and shape was proposed for handling bosom and prostate tumor pictures [45]. A successive issue in cores division is the partition of cores among gatherings of firmly pressed cells [46–48]. Another regular issue is the partition of a structure-featuring stain, for example, hematoxylin and eosin into subcellular segments. In these pictures, structures such as cores, cytoplasm, and blood are portrayed by shading modes that differ somewhat between tests, making measurable displaying a well-known approach [49]. The mean-move calculation joins both shading and spatial territory into a strong system that recognizes these modes in a changed shading space [50].

ii) **Region Segmentation:** Region division can expect numerous structures, from the division of basic multi-cellular structures to totally unsupervised segmentation of tissues. Due to the textural appearance of numerous tissues or locales, surface division strategies from the more extensive picture examination group are often adjusted to pathology imaging [78]. Locale division is commonly preceded by a succession of activities, including highlight extraction, to make a middle of the road portrayal of picture information that catches surface separating data. Multiresolution techniques in view of filter banks are usually used to abuse the presence of tissues at different scales [51, 52]. Different strategies depend on spatial measurements and factual geometry [53, 54]. The two-point connection work measures the spatial dispersions of tissue segments, for example, cores and cytoplasms, shaping a spatial–factual mark for each tissue composition. These highlights have the preferred standpoint that they can be registered deterministically in a computationally effective manner [55].

iii) **Computer-Aided Diagnosis:** Computer-helped finding (CAD) is the most dynamic research territory in pathology picture investigation. The point of these frameworks is to decrease inconstancy and blunder in conclusion by copying built up analytic techniques. There are various frameworks produced for an extensive range of illnesses, including cervical disease [56], prostate malignancy [57–59], bosom growth [60–63], colon tumor [64], neuroblastoma [65–67], and follicular lymphoma [68–70]. CAD frameworks normally contain numerous modules executing object division, locale division, and highlight extraction to accomplish characterization of sickness. A short review of two CAD frameworks is introduced underneath to show basic CAD ideas.

iv) **Neuroblastoma CAD System:** Neuroblastoma (NB) is a standout amongst the most frequent tumors of the sensory system in youngsters. The procedure of NB analysis is directed by profoundly concentrated pathologists, is tedious, and is inclined to inconstancy. Heterogeneous tumors show a specific issue since the pathologist is restricted in the number of fields they can for all intents and purposes survey amid analysis. Two segments of NB reviewing were robotized, keeping in mind the end goal to beat challenges posed by the heterogeneity of these tumors [65, 67, 71]. Stromal arrangement utilizes a surface investigation to recognize locales as either stroma-rich or stroma-poor. The level of differentiation is assessed by portioning the picture into cores and cytoplasm parts and breaking down the surface made by these segments. The two investigations are performed in a multiresolution way to upgrade calculation and precision.

v) **Lymphoma CAD System:** Follicular lymphoma (FL) is the second most basic type of non-Hodgkin's lymphoma. Beginning inside follicle-like multicellular areas, FL includes primarily all the different lymphomas, which are evaluated using follicles agent, first utilizing low magnification to distinguish follicles, and after that a 40X target to look for centroblasts. In heterogeneous tumors, the centroblast check can change broadly starting with one follicle and then going onto the next, leaving evaluating subject to considerable fluctuation.

An electronic framework was created to mechanize the centroblast checking, utilizing a crossbreed enlistment/classification approach that incorporates data from various stains [68–70]. Follicle areas are hard to recognize in a structure featuring stain, thus follicles are first distinguished from counteracting agent recolor segments utilizing district division. Centroblasts are hard to recognize from other cell compositions in the counteracting agent stain, thus the follicle limits are mapped utilizing a non-rigid enlistment to an adjoining bit of tissue recolor with a structure featuring stain. Cells in the mapped follicle locales are then fragmented and grouped.

9.4.1 Demand for Scale

Calculation, stockpiling, and systems administration remain significant challenges for slide imaging. Present-day commercial scanners are equipped to create pictures at 40X target amplification, and are developing ever quicker. At 40X amplification, the digitization of a solitary 2 cm² test contains 7.5 billion pixels (21 GB uncompressed). At this scale, in-center examinations on a solitary machine are impractical. Superior registering has been utilized in both research and business pathology picture investigation applications to address the scale challenge. The rise of repurposed product illustrations equipment for general purpose figuring has been a promising advancement, offering equipment increasing the speed of picture examination on work area frameworks in situations where processing groups are not accessible. A few organizations at present offer some type of parallel figuring as an element in their picture examination bundle lineups. Past the test of crude picture stockpiling there is a necessity to record algorithmic outcomes and demonstrate them in an accessible shape. This theme is further discussed in Section 9.5.

9.5 Management Infrastructure

The foundation for analyzing, overseeing, questioning, and sharing WSI is basic for its utilization in electronic human services records. Configuration should bolster both fundamental obtaining and the administration and trade of WSI and inferred scientific outcomes.

9.5.1 WSI Acquisition, Management, and Exchange

The real difficulties for WSI administration originate from the speed of procurement, the size of information created, the variety of picture arrangements, and the directions encompassing data innovation utilized for care. All together for a pathology practice to digitize all slides, instruments should have the capacity to check WSI at high amplification (40X goal) at normal speeds close to that plotted in Figure 9.3. These normal velocities must record for the variable measure of surface region secured by tissue areas: a run of the mill segment measures 15 mm × 15 mm (225 mm^2); however, segments can go anyplace from a couple of millimeters to an extreme of 25 mm × 55 mm (1,375 mm^2, more than six times the normal segment's surface region). At 2 m for a 15 mm × 15 mm segment at 20X target amplification, even the normal pathology hone requires a few full-time scanners to accomplish digitization. A scanner with significant enough speed and accuracy will be required, keeping in mind the end goal to achieve this objective. Various conventions have been produced to oversee and trade social insurance data, including the digital image communications in medicine (DICOM) [72–74] in radiology and Health Level Seven (HL7) [18, 73, 74] for clinical information.

Slide Density	No Of Slides Per Day	Time Consumed Per Slide
Average slides	330	4.36
Large slides	1350	1.06

FIGURE 9.3
Slide production estimates. An average-sized pathology practice can reasonably produce 80,000 slides per year.

Classification strategies have likewise been created to build up a most widely used language for the phrasing in information trade, including logical observation identifier names and codes (LOINC) [32] and unified medical language system (UMLS). Pathology imaging poses special necessities related to overseeing and trading expansive pictures and executing complex inquiries over accumulations of pictures and determined information. Advancement of norms for WSI is a generally new and quickly developing field. The Open Microscopy Environment (OME) venture has built up an information model and administration system that can be utilized to speak to, trade, and oversee microscopy picture information and metadata [75]. The DICOM working gathering for pathology, known as Working Group 26, as of late created two supplements, 122 and 145, for formal portrayals for examples and WSI pictures [71,72]. Supplement 145 characterizes a tiled multiresolution portrayal for quick recovery and survey. A DICOM-based portrayal offers the benefits of a general standard that is additionally good with existing picture archiving and communication systems (PACS) utilized in radiology offices. Figure 9.3 demonstrates the slide production per day in a pathology lab.

9.5.2 Pathology Analytical Imaging Infrastructure

Pathology pictures are frequently connected with rich meta-information, including explanations made by people, markups, highlights, and characterizations processed from robotized picture calculations. The precise examination of WSI brings about huge measures of morphological data at different bio-sensible scales. The data produced by this procedure has huge potential for giving understanding with respect to the beginning and progression of infection. Noteworthy deterrents that have a tendency to diminish more extensive appropriation of these new advances all through the clinical and established researchers are administration, inquiry, and joining of this metadata. The Pathology Analytic Imaging Standards (PAIS) venture builds up far-reaching information for virtual slide–related pictures, explanations, markups, and highlighted data [76].

PAIS likewise produces an information administration infrastructure, basically a medicinal imaging geological data framework (GIS), to help inquiries for information recovery in view of examination and picture metadata, questions for correlation of results from various investigations, and spatial questions to evaluate relative pervasiveness of highlights and arranged protests and to recover accumulations of divided locales and highlights. With the size of information (~2 GB metadata per WSI), elite parallel database design is basic to help such inquiries and scale to huge accumulations of WSIs. Complex inquiries, for example, calculation result examination crosswise over WSI, are serious about information and calculation. Support of fast reaction of these inquiries requires particular questioning calculations executing on superior computing framework.

9.6 Investigation in Computerized Pathology Medical Imaging

Digitization and picture investigation are required to essentially modify the field of pathology, as digitization equipment and business picture examination devices multiply in the clinical circle. The vast majority of the business picture analysis instruments are used for expanding throughput and lessening fluctuation. A few investigation apparatuses as of now have an endorsement from the FDA for mechanizing routine scoring of IHC-recolor slides. The examination group has to a great extent followed this pattern, with attention to frameworks that copy analytic strategies. However, following an evaluation plan is more difficult than performing protein analysis due to the rules. In this segment, we consider how picture investigation may use the rising wealth of digitized pathology to surpass copying of human pathologists to enhance prognosis, therapeutics, and comprehension of complex illnesses like a tumor. The combination of patient registries, doctor's facility data frameworks, and advanced pathology archives will give an examination test where imaging and molecular information can be drilled to connect persistent genomes with morphology. This will allow clinicians and researchers to think about natural attributes crosswise over associates of subjects to make new arrangements of patient populaces to better drive personalization of treatment. Below are some techniques for segmenting abnormal pathological images.

9.6.1 Nephropathy Glomerulosclerosis: Integrative Using Active Contour

This is a proposed division strategy for distinguishing the adjustments in the glomerulus caused in patients experiencing diabetic illness [77]. We propose utilizing the Chan-Vese display for dynamic shapes, which is an adaptable and effective strategy and ready to fragment different classifications of pictures. With a variety in the parameters' estimations of Chan-Vese display, a better yield is obtained [79–82, 85]. A sample of two segmented images is mentioned in Fig 9.4.

9.6.2 Mapping Molecules in the Tumor Microenvironment Using k-Means Clustering

Tumors are like any other organ with different situations satisfying distinctive parts [83, 84]. To exhibit the execution of the proposed approach, we have taken RCC pictures. As a contextual analysis, the introduced method is evaluated on the affected area of RCC image. We have taken a cancer disease called renal cell carcinoma (RCC) for detecting and segmenting the affected area of RCC. Figure 9.5 shows some images of the affected RCC segmented are using the proposed method [87–92].

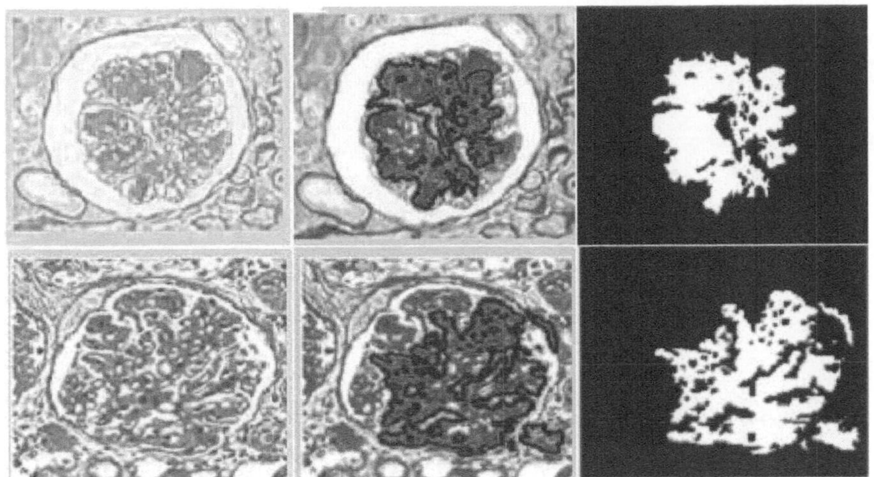

FIGURE 9.4
Counter-based segmentation of glomerular images.

The nearness of a lot of varieties in the informational collection makes it more reasonable. Figure 9.5 demonstrates the division on the sequence of an influenced range of RCC utilizing the k-means clustering system. We have divided the data of the image into four groups as shown in Figure 9.5 and obviously the fourth cluster is an accurate segmentation of the RCC image. Through the experiential perceptions, it was found that utilizing three or four cluster yields a great division is performed. In this way, in this investigation input, images are picked up into four sections according to their prerequisites.

9.6.3 Challenges Transversely Occurred in the Process

The advancements in transmission technology, digitization, and analysis have significantly reduced the burden of computation and information delivery. The big data from picture examination have superior and adaptability prerequisites comparable with big business medicinal services information, yet exhibit one of kind of difficulties. Later on, even medium-scale healing facilities and research activities will require the ability to oversee a huge number of high-determination pictures, execute and oversee interrelated examination pipe-lines, and question trillions of minuscule items and their highlights. These applications request quick stacking and question reactions, and additionally decisive inquiry interfaces for ease of use.

On the computational front, the profound information parallel nature of pathology picture examination gives a chance to both apply new equipment

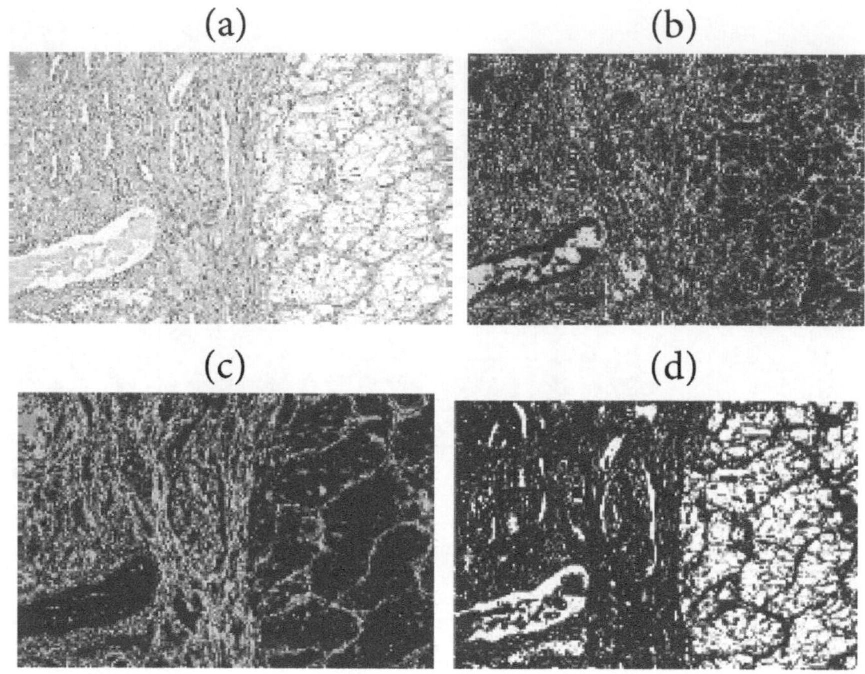

FIGURE 9.5
k-means clustering for patient infected with RCC disease (a) original image, (b) first cluster, (c) second cluster, (d) third cluster image is the final output where the tumor area i.e. affected area is clearly segmented.

and program propels. Universally useful illustrations handling units (GPGPUs) have raised as a well-known execution stage for some information concentrated computational science applications. Heterogeneous elite registering arrangements comprising of multicore CPUs and different GPGPUs are getting to be normal, giving an appealing contrasting option to more customary homogeneous processing bunches. Such heterogeneous frameworks offer huge figuring capability at sensible obtaining and working expenses, providing singular specialists with the way to investigate distinctive examination procedures at important information scales. Thus, framework programming stacks created for information parallel venture applications, for example, Map Reduce [86], can be utilized to give adaptable, proficient, practical answers for pathology picture investigation. Broad adoption of these new innovations brings another arrangement of challenges, nonetheless: GPGPUs include another level of many-sided quality to application memory pecking orders, and new apparatuses must be produced to characterize, send, and oversee circulated computations crosswise over heterogeneous frameworks and systems.

Capacity advances have likewise progressed fundamentally in the previous decade. Storage devices like solid state drive (SSD) currently have low storage capacity compared to other contemporary storage devices. We can expect that a capacity framework made out of various levels of coupled turning drives and SSDs in RAID designs will turn out to be more typical. In such setups, high limit varieties of customary circles would give longer-term stockpiling and rapid access for tasks that are portrayed by consecutive information (e.g. spilling of picture information for examination). In any case, new storage, ordering, information arranging methods, and programming segments will be expected to exploit these numerous levels of capacity frameworks.

There have been generous advances in arranging switches and organizing conventions for intra-cluster communications. Advances, for example Infiniband, give low-idleness, high-transmission capacity correspondence substrates. Nonetheless, advances in wide-region organizing have been generally moderate. Indeed, even as multi-gigabit systems are becoming all the more broadly conveyed inside foundations, and strategies for giving acceptably high successful bandwidths for appropriated applications have been produced, low-dormancy access to remote assets remains an issue on product wide-territory systems. Proficient compression, dynamic information transmission, and insightful information are reserving and calculation reuse techniques that will keep on playing basic parts in empowering computerized pathology and logical coordinated efforts, including vast pathology picture informational indexes.

9.7 Conclusion

The capacity to quantitatively describe infection classification and process from numerous natural scales and measurements can possibly empower advancement of preventive methodologies and restorative medications that are decisively focused on every individual patient, otherwise called customized prescription. The advances in pathology imaging innovations discussed in this paper are setting up the therapeutic expert's capacity to quickly catch and endeavor tremendous measures of multi-stage, multi-dimensional information on every patient's hereditary foundation, organic capacity, and structure. High determination and high-throughput instruments are being employed routinely in medicinal science, as well as in human services conveyance settings at a quickening rate. As this decade advances, critical advances in restorative data innovations will change extensive volumes of multi-scale, multi-dimensional information into significant data to drive the revelation, improvement, and conveyance of new systems of averting, diagnosing, and recuperating complex illnesses. Inflow rehearses, atomic information, and human-produced pathology elucidations are utilized to create medicines for an ailment.

Previously, pathologists characterized issues by physically perceived examples. Later, numerous scientists have shown that now and again pathological picture investigation will benefit the appropriate classification of unrecognized cases. The examination and data administration strategies depicted in this paper discuss the innovation that will be utilized to give profoundly focused and customized medicinal services in the following decade.

References

1. R. S. Weinstein, "Prospects for telepathology," *Human Pathology*, vol. 17, no. 5, pp. 433–434, 1986.
2. R. S. Weinstein, K. J. Bloom, and L. S. Rozek, "Telepathology and the networking of pathology diagnostic services," *Archives of Pathology & Laboratory Medicine*, vol. 111, no. 7, pp. 646–652, 1987.
3. CAPV Transforming Pathologists, May 15, 2011 [Online]. Available at: http://www.cap.org/apps/docs/membership/transformation/new/index.html.
4. Description of Examinations V Primary, May 15, 2011 [Online]. Available at: http://www.abpath.org/DescriptionOfExamsAPCP.pdf.
5. N. Gray, "Milestones in light microscopy," *Nature Cell Biology*, vol. 11, no. 10, p. 1165, 2009.
6. R. Hooke, *Micrographia: Or Some Physiological Descriptions of Minute Bodies, Made by Magnifying Glasses with Observations and Inquiries Thereupon*, John Martyn and James Allestry, London, UK, 1665.
7. D. Evanko, A. Heinrichs, and C. K. Rosenthal, "Milestones in light microscopy," *Nature*, May 14, 2011 [Online]. Available at: http://www.nature.com/milestones/milelight/index.html.
8. G. A. Meijer, J. A. Belen, P. J. van Diest, and J. P. Baak, "Origins of ... image analysis in clinical pathology," *Journal of Clinical Pathology*, vol. 50, no. 5, pp. 365–370, 1997.
9. A. Bayer, "Ueber eine neue Klasse von Farbstoffen," *Berichte der deutschen chemischen Gesellshaft*, vol. 4, pp. 555–558, 1871.
10. E. Abbe, "Beitrage zur Theorie des Mikroscops und der Mikroskopischen Wahrnehmung," *Archiv fur Mikroskopische Anatomie*, vol. 9, pp. 413–418, 1873.
11. O. Heimstadt, "Das Fluoreszenzmikroskop," *Z Wiss Mikrosk*, vol. 28, pp. 330–337, 1911.
12. A. H. Coons, "The beginnings of immunofluorescence," *Journal of Immunology*, vol. 87, pp. 499–503, 1961.
13. A. H. Coons, H. J. Creech, and R. N. Jones, "Immunological properties of an antibody containing a fluorescent group," *Proceedings of the Society for Experimental Biology and Medicine*, vol. 47, pp. 200–202, 1941.

14. J. W. Lichtman and J. A. Conchello, "Fluorescence microscopy," *Nature Methods*, vol. 2, no. 12, pp. 910–919, 2005.
15. R. C. Graham Jr. and M. J. Karnovsky, "The early stages of absorption of injected horseradish peroxidase in the proximal tubules of mouse kidney: ultrastructural cytochemistry by a new technique," *Journal of Histochemistry & Cytochemistry*, vol. 14, no. 4, pp. 291–302, 1966.
16. M. J. Karnovsky, "BA pathologist's odyssey," *Annual Review of Pathology*, vol. 1, pp. 1–22, 2006.
17. T. S. Reese and M. J. Karnovsky, "Fine structural localization of a blood-brain barrier to exogenous peroxidase," *Journal of Cell Biology*, vol. 34, no. 1, pp. 207–217, 1967.
18. W. Straus, "Segregation of an intravenously injected protein by droplets of the cells of rat kidneys," *Journal of Biophysical and Biochemical Cytology*, vol. 3, no. 6, pp. 1037–1040, 1957.
19. M. Chalfie, Y. Tu, G. Euskirchen, W. W. Ward, and D. C. Prasher, "Green fluorescent protein as a marker for gene expression," *Science*, vol. 263, no. 5148, pp. 802–805, 1994.
20. R. Y. Tsien, "The green fluorescent protein," *Annual Review of Biochemistry*, vol. 67, pp. 509–544, 1998.
21. A. H. Mayall and M. L. Mendelsohn, "Deoxyribonucleic acid cytophotometry of stained human leukocytes V Part II: The mechanical scanner CYDAC, the theory of scanning photometry and the magnitude of residual errors," *Journal of Histochemistry & Cytochemistry*, vol. 18, no. 6, pp. 383–407, 1970.
22. P. Nipkow, German Patent 30 105, 1885.
23. T. J. Eide, I. Nordrum, and H. Stalsberg, "The validity of frozen section diagnosis based on video-microscopy," *Zentralblatt für Pathologie*, vol. 138, no. 6, pp. 405–407, 1992.
24. T. J. Eide and I. Nordrum, "Frozen section service via the telenetwork in northern Norway," *Zentralblatt für Pathologie*, vol. 138, no. 6, pp. 409–412, 1992.
25. K. J. Kaplan, J. R. Burgess, G. D. Sandberg, C. P. Myers, T. R. Bigott, and R. B. Greenspan, "Use of robotic telepathology for frozen-section diagnosis: a retrospective trial of a telepathology system for intraoperative consultation," *Modern Pathology*, vol. 15, no. 11, pp. 1197–1204, 2002.
26. I. Nordrum, B. Engum, E. Rinde, A. Finseth,H. Ericsson, M. Kearney, H. Stalsberg, and T. J. Eide, "Remote frozen section service: a telepathology project in northern Norway," *Human Pathology*, vol. 22, no. 6, pp. 514–518, 1991.
27. E. G. Fey and S. Penman, "The morphological oncogenic signature. Reorganization of epithelial cytoarchitecture and metabolic regulation by tumor promoters and by transformation," *Developmental Biology*, vol. 3, pp. 81–100, 1986.
28. R. S. Weinstein, A. R. Graham, L. C. Richter, G. P. Barker, E. A. Krupinski, A. M. Lopez, K. A. Erps, A. K. Bhattacharyya, Y. Yagi, and J. R. Gilbertson, "Overview of telepathology, virtual microscopy, and whole slide imaging: prospects for the future," *Human Pathology*, vol. 40, no. 8, pp. 1057–1069, 2009.
29. S. Williams, W. H. Henricks, M. J. Becich, M. Toscano, and A. B. Carter, "Telepathology for patient care: what am I getting myself into?" *Advances in Anatomic Pathology*, vol. 17, no. 2, pp. 130–149, 2010.

30. A. Afework, M. D. Beynon, F. Bustamante, S. Cho, A. Demarzo, R. Ferreira, R. Miller, M. Silberman, J. Saltz, A. Sussman, and H. Tsang, "Digital dynamic telepathology. The virtual microscope," in *Proceedings of the AMIA Symposium*, 1998, pp. 912–916.
31. R. Ferreira, B. Moon, J. Humphries, A. Sussman, J. Saltz, R. Miller, and A. Demarzo, "The virtual microscope," in *Proceedings of the AMIA Symposium*, 1997, pp. 449–453.
32. A. J. McDonald, S. M. Huff, J. G. Suico, G. Hill, D. Leavelle, R. Aller, A. Forrey, K. Mercer, G. DeMoor, J. Hook, W. Williams, J. Case, and P. Maloney, "BLOINC, a universal standard for identifying laboratory observations: a 5-year update," *Clinical Chemistry*, vol. 49, no. 4, pp. 624–633, 2003.
33. M. G. Rojo, G. B. Garcia, C. P. Mateos, J. G. Garcia, and M. C. Vicente, "Critical comparison of 31 commercially available digital slide systems in pathology," *International Journal of Surgical Pathology*, vol. 14, no. 4, pp. 285–305, 2006.
34. A. C. Wilbur *et al.* "Whole-slide imaging digital pathology as a platform for teleconsultation: a pilot study using paired subspecialist correlations," *Archives of Pathology & Laboratory Medicine*, vol. 133, no. 12, pp. 1949–1953, 2009.
35. J. R. Gilbertson *et al.* "Primary histological diagnosis using automated whole slide imaging: a validation study," *BMC Clinical Pathology*, vol. 6, p. 4, 2006.
36. U. J. Balis, R. D. Aller, and E. R. AshWood, "Informatics training in U.S. pathology residency programs. Results of a survey," *American Journal of Clinical Pathology*, vol. 100, no. 4 Supplement 1, pp. S44–S47, 1993.
37. J. H. Harrison Jr., "Pathology informatics questions and answers from the University of Pittsburgh pathology residency informatics rotation," *Archives of Pathology & Laboratory Medicine*, vol. 128, no. 1, pp. 71–83, 2004.
38. W. H. Hendricks *et al.*, "Informatics training in pathology residency programs: proposed learning objectives and skill sets for the new millennium," *Archives of Pathology & Laboratory Medicine*, vol. 127, no. 8, pp. 1009–1018, 2003.
39. C. Safran *et al.*, "Program requirements for fellowship education in the subspecialty of clinical informatics," *Journal of the American Medical Informatics Association*, vol. 16, no. 2, pp. 158–616, 2009.
40. R. M. Gardner *et al.*, "Core content for the subspecialty of clinical informatics," *Journal of the American Medical Informatics Association*, vol. 16, no. 2, pp. 153–157, 2009.
41. Charité Universitätsmedizin Berlin, European Scanner Contest, May 2011 [Online]. Available at: http://scanner-contest.charite.de
42. A. C. Ruifrok, R. L. Katz, and D. A. Johnston, "Comparison of quantification of histochemical staining by hue-saturation-intensity (HSI) transformation and color-deconvolution," *Applied Immunohistochemistry & Molecular Morphology*, vol. 11, no. 1, pp. 85–91, 2003.
43. C. Cantaloni, R. E. Tonini, C. Eccher, L. Morelli, E. Leonardi, E. Bragantini, D. Aldovini, S. Fasanella, A. Ferro, D. Cazzolli, G. Berlanda, P. Dalla Palma, and M. Barbareschi, "Diagnostic value of automated Her2 evaluation in breast cancer: a study on 272 equivocal (score 2+) Her2 immunoreactive cases using an FDA approved system," *Applied Immunohistochemistry & Molecular Morphology*, vol. 19, no. 4, pp. 306–312, 2011.
44. M. N. Gurcan, L. E. Boucheron, A. Can, A. Madabhushi, N. M. Rajpoot, and B. Yener, "Histopathological image analysis: a review," *IEEE Reviews in Biomedical Engineering*, vol. 2, pp. 147–171, 2009.

45. S. Naik, S. Doyle, S. Agner, A. Madabhushi, M. Feldman, and J. Tomaszewski, "Automated gland and nuclei segmentation for grading of prostate and breast cancer histopathology," in *Proceedings/IEEE International Symposium on Biomedical Imaging*, 2008, pp. 284–287.
46. H. Fatakdawala, J. Xu, A. Basavanhally, G. Bhanot, S. Ganesan, M. Feldman, J. E. Tomaszewski, and A. Madabhushi, "Expectation-maximization-driven geodesic active contour with overlap resolution (EMaGACOR): application to lymphocyte segmentation on breast cancer histopathology," *IEEE Transactions on Biomedical Engineering*, vol. 57, no. 7, pp. 1676–1689, 2010.
47. G. Li, T. Liu, J. Nie, L. Guo, J. Chen, J. Zhu, W. Xia, A. Mara, S. Holley, and S. T. Wong, "Segmentation of touching cell nuclei using gradient flow tracking," *Journal of Microscopy*, vol. 231, no. 1, pp. 47–58, 2008.
48. Q. Wen, H. Chang, and B. Parvin, "A Delaunay triangulation approach for segmenting clumps of nuclei," in *Proceedings of the Sixth IEEE International Conference on Symposium on Biomedical Imaging: From Nano to Macro*, Boston, MA, 2009, pp. 9–12.
49. M. N. Gurcan, J. Kong, O. Sertel, B. B. Cambazoglu, J. Saltz, and U. Catalyurek, "Computerized pathological image analysis for neuroblastoma prognosis," in *Proceedings of the AMIA Annual Symposium*, 2007, pp. 304–308.
50. A. Comaniciu and P. Meer, "Cell image segmentation for diagnostic pathology," in *Advanced Algorithmic Approaches to Medical Image Segmentation*, Springer-Verlag, New York, 2002, pp. 541–558.
51. J. Han, H. Chang, L. Loss, Z. Kai, F. L. Baehner, J. W. Gray, P. Spellman, and B. Parvin, "Comparison of sparse coding and kernel methods for histopathological classification of glioblastoma multiforme," in *Proceedings of the Sixth IEEE International Conference on Symposium on Biomedical Imaging: From Nano to Macro*, 2011, pp. 711–714.
52. N. Signolle, M. Revenu, B. Plancoulaine, and P. Herlin, "Wavelet-based multiscale texture segmentation: application to stromal compartment characterization on virtual slides," *Signal Processing*, vol. 90, no. 8, pp. 2412–2422, 2010.
53. J. Hipp, J. Cheng, J. C. Hanson, W. Yan, P. Taylor, N. Hu, J. Rodriguez-Canales, J. Hipp, M. A. Tangrea, M. R. Emmert-Buck, and U. Balis, "SIVQ-aided laser capture microdissection: a tool for high-throughput expression profiling," *Journal of Pathology Informatics*, vol. 2, no. 19, 2011. DOI: 10.4103/2153-3539.78500.
54. K. Mosaliganti *et al.*, "Tensor classification of N-point correlation function features for histology tissue segmentation," *Medical Image Analysis*, vol. 13, no. 1, pp. 156–166, 2009.
55. L. A. D. Cooper, J. Saltz, R. Machiraju, and H. Kun, "Two-point correlation as a feature for histology images: feature space structure and correlation updating," *Proceedings of the IEEE Computer Society Conference on Computer Vision and Pattern Recognition Workshops*, 2010, pp. 79–86.
56. S. J. Keenan, J. Diamond, W. G. McCluggage, H. Bharucha, D. Thompson, P. H. Bartels, and P. W. Hamilton, "An automated machine vision system for the histological grading of cervical intraepithelial neoplasia (CIN)," *Journal of Pathology*, vol. 192, no. 3, pp. 351–662, 2000.
57. A. Tabesh, M. Teverovskiy, H.-Y. Pang, V. P. Kumar, D. Verbel, A. Kotsianti, and O. Saidi, "Multifeature prostate cancer diagnosis and Gleason grading of histological images," *IEEE Transactions on Medical Imaging*, vol. 26, no. 10, pp. 1366–1378, 2007.

58. P. Khurd, C. Bahlmann, P. Maday, A. Kamen, S. Gibbs-Strauss, E. M. Genega, and J. V. Frangioni, "Computer-aided Gleason grading of prostate cancer histopathological images using texton forests," in *Proceedings/IEEE International Symposium on Biomedical Imaging*, 2010, pp. 636–639.
59. E. Yu, J. P. Monaco, J. Tomaszewski, N. Shih, M. Feldman, and A. Madabhushi, "High-throughput detection of prostate cancer in histological sections using probabilistic pairwise Markov models," *Medical Image Analysis*, vol. 14, no. 4, pp. 617–629, 2010.
60. A. N. Basavanhally, S. Ganesan, S. Agner, J. P. Monaco, M. D. Feldman, J. E. Tomaszewski, G. Bhanot, and A. Madabhushi, "Computerized image-based detection and grading of lymphocytic infiltration in HER2 plus breast cancer histopathology," *IEEE Transactions on Biomedical Engineering*, vol. 57, no. 3, pp. 642–653, 2010.
61. A. Madabhushi, S. Agner, A. Basavanhally, S. Doyle, and G. Lee, "Computer-aided prognosis: predicting patient and disease outcome via quantitative fusion of multi-scale, multi-modal data," *Computerized Medical Imaging and Graphics*, vol. 35, no. 7–8, pp. 506–514, 2011.
62. G. Van De Wouwer, B. Weyn, P. Scheunders, W. Jacob, E. Van Marck, and D. Van Dyck, "Wavelets as chromatin texture descriptors for the automated identification of neoplastic nuclei," *Journal of Microscopy*, vol. 197, no. 1, pp. 25–35, 2000.
63. B. Weyn, G. van de Wouwer, A. van Daele, P. Scheunders, D. van Dyck, E. Van Marck, and W. Jacob, "Automated breast tumor diagnosis and grading based on wavelet chromatin texture description," *Cytometry*, vol. 33, no. 1, pp. 32–40, 1998.
64. N. Rajpoot and K. Masood, "Texture based classification of hyperspectral colon biopsy samples using CBLP," in *Proceedings of the ISBI*, 2009, pp. 1011–1014. DOI: 10.1109/ISBI.2009.5193226.
65. K. L. Boyer, J. H. Saltz, and M. N. Gurcan, J. Kong, and O. Sertel, "A multi-resolution image analysis system for computer-assisted grading of neuroblastoma differentiation," in *Proceedings of the SPIE*, 2008, vol. 6915, no. 69151T.
66. O. Sertel, U. V. Catalyurek, H. Shimada, and M. N. Gurcan, "Computer-aided prognosis of neuroblastoma: detection of mitosis and karyorrhexis cells in digitized histological images," in *Proceedings of the IEEE Conference on Engineering in Medicine and Biology Society*, 2009, vol. 2009, pp. 1433–1436.
67. M. Pennell, G. Lozanski, K. Belkacem-Boussaid, A. Shana'ah, and M. Gurcan, "Computer-aided classification of centroblast cells in follicular lymphoma," *Analytical and Quantitative Cytology and Histology*, vol. 32, no. 5, pp. 254–260, 2010.
68. H. Shimada, U. Catalyurek, O. Sertel, J. Kong, J. H. Saltz, and M. N. Gurcan, "Computer-aided prognosis of neuroblastoma on whole-slide images: classification of stromal development," *Pattern Recognition*, vol. 42, no. 6, pp. 1093–1103, 2009.
69. L. Cooper, O. Sertel, J. Kong, G. Lozanski, K. Huang, and M. Gurcan, "Feature-based registration of histopathology images with different stains: an application for computerized follicular lymphoma prognosis," *Computer Methods and Programs in Biomedicine*, vol. 96, no. 3, pp. 182–192, 2009.
70. O. Sertel, G. Lozanski, A. Shana'ah, and M. N. Gurcan, "Computer-aided detection of centroblasts for follicular lymphoma grading using adaptive likelihood-based cell segmentation," *IEEE Transactions on Biomedical Engineering*, vol. 57, no. 10, pp. 2613–2616, 2010.

71. K. Jun, O. Sertel, H. Shimada, K. Boyer, J. Saltz, and M. Gurcan, "Computer-aided grading of neuroblastic differentiation: multi-resolution and multi-classifier approach," *Proceedings of the IEEE International Conference on Image Processing*, 2007, pp. V-525–V-528.
72. DICOM, Digital imaging and communications in medicine, May 2011 [Online]. Available at: http://medical.nema.org/.
73. C. Daniel, F. Macary, M. Garcia-Rojo, J. Klossa, A. Laurinavičius, B. A. Beckwith, and V. Della Mae, "Recent advances in standards for collaborative digital anatomic pathology," *Diagnostic Pathology*, vol. 6, no. Suppl 1, p. S17, 2011.
74. C. Daniel, M. G. Rojo, J. Klossa, V. Della Mea, D. Booker, B. A. Beckwith, and T. Schrader, "Standardizing the use of whole slide images in digital pathology," *Computerized Medical Imaging and Graphics*, vol. 35, no. 7–8, pp. 496–505, 2011.
75. I. G. Goldberg, C. Allan, J. M. Burel, D. Creager, A. Falconi, H. Hochheiser, J. Johnston, J. Mellen, P. K. Sorger, and J. R. Swedlow, "The open microscopy environment (OME) data model and XML file: open tools for informatics and quantitative analysis in biological imaging," *Genome Biology*, vol. 6, no. 5, p. R47, 2005.
76. D. J. Foran, L. Yang, W. Chen, J. Hu, L. A. Goodell, M. Reiss, F. Wang, T. Kurc, T. Pan, A. Sharma, and J. H. Saltz, "Image Miner: A software system for comparative analysis of tissue microarrays using content-based image retrieval, high-performance computing, and grid technology," *Journal of the American Medical Informatics Association*, vol. 18, no. 4, pp. 403–415, 2011.
77. M. Ravi and Ravindra S. Hegadi, "Detection of glomerulosclerosis in diabetic nephropathy using contour-based segmentation," *Procedia Computer Science*, vol. 45, pp. 244–249, 2015.
78. R. Deriche and N. Paragios, "Geodesic active regions and level set methods for supervised texture segmentation," *International Journal of Computer Vision*, 2002.
79. D. Mumford and J. Shah, "Optimal approximation by piecewise smooth functions and associated variation problems," *Communications on Pure and Applied Mathematics*, vol. 42, pp. 577–685, 1989.
80. Olivier Rousseau and Yves Bourgault, *Heart Segmentation with an Iterative Chan-Ve Algorithm*, University of Ottawa, Ontario, ON, 2009.
81. P. Kleihues, P. C. Burger, K. Aldape, D. J. Brat, W. Biernat, D. D. Bigner, Y. Nakazato, K. H. Plate, F. Ginagaspero, A. von Deimling, H. Ohgaki, O. D. Weistler, and W. K. Cevenee, *WHO Classification of Tumors of the Central Nervous System*, 4th ed. IARC Press, Lyon, France, 2007.
82. L. A. Cooper, J. Kong, D. A. Gutman, F. Wang, S. R. Cholleti, T. C. Pan, P. M. Widener, A. Sharma, T. Mikkelsen, A. E. Flanders, D. L. Rubin, E. G. Van Meir, T. M. Kurc, C. S. Moreno, D. J. Brat, and J. H. Saltz, "An integrative approach for in silico glioma research," *IEEE Transactions on Biomedical Engineering*, vol. 57, no. 10, pp. 2617–2621, 2010.
83. D. Hanahan and R. A. Weinberg, "The hallmarks of cancer," *Cell*, vol. 100, no. 1, pp. 57–70, 2000.
84. D. Hanahan and R. A. Weinberg, "Hallmarks of cancer: the next generation," *Cell*, vol. 144, no. 5, pp. 646–774, 2011.
85. W. R. Schwartz and H. Pedrini, "Segmentation of satellite image by wavelet transforms based on color and texture features," in *4th ISA in Visual Computing*, 2008, pp. 113–122.

86. S. J. Wang and H. C. Chen, "Difference-based of visible color for quantitative evaluation of colour segmentation," *IEEE Vision Image and Signal Processing*, vol. 153, no. 5, pp. 598–609, 2006.
87. N. R. Pal and S. K. Pal, "A review on image segmentation techniques," *Pattern Recognition*, vol. 26, no. 9, pp. 1227–1294, 1993.
88. Y. Boykov, "Graph cuts and efficient N-D image segmentation," *International Journal of Computer Vision (IJCV)*, vol. 70, no. 2, pp. 109–131, 2006.
89. R. Krishnapuram and H. Frigui, "Competitive clustering by agglomeration," *Pattern Recognition*, vol. 30, no. 7, pp. 1109–1119, 1997.
90. Gullanar M. Hadi and Nassir H. Salman, "Image segmentation based on single seed region growing algorithm," in *1st International Conference on Engineering and Innovative Technology*, 2016.
91. Lee A. D. Cooper, Jun Kong, and David A. Gutman, Digital pathology data-intensive frontier in medical imaging, *Proceedings of the IEEE*, 2012.
92. M. Ravi and Ravindra S. Hegadi, "Detection of renal cell carcinoma – a kidney cancer using K-means clustering segmentation focused on the pathological microscopic image," *Journal of Advanced Research in Dynamical and Control Systems*, no. 05-Special Issue, pp. 144–149, 2017.
93. D. D. Ruikar, R. S. Hegadi, and K. C. Santosh, Contrast stretching-based unwanted artifacts removal from CT images in recent trends in image processing and pattern recognition (accepted). Springer, 2019.
94. S. Vajda and K. C. Santosh, "A fast k-nearest neighbor classifier using unsupervised clustering," in *International Conference on Recent Trends in Image Processing and Pattern Recognition*, 2016, pp. 185–193, Springer, Singapore.
95. A. Karargyris, J. Siegelman, D. Tzortzis, S. Jaeger, S. Candemir, Z. Xue, K. C. Santosh, S. Vajda, S. Antani, L. Folio, and G. R. Thoma, "Combination of texture and shape features to detect pulmonary abnormalities in digital chest X-rays," *International Journal of Computer Assisted Radiology and Surgery*, vol. 11, no. 1, pp. 99–106, 2016.
96. K. C. Santosh, L. Wendling, S. Antani, and G. R. Thoma, "Overlaid arrow detection for labeling regions of interest in biomedical images," *IEEE Intelligent Systems*, vol. 31, no. 3, pp. 66–75, 2016.
97. K. C. Santosh, S. Candemir, S. Jäger, L. Folio, A. Karargyris, S. Antani, and G. Thoma, "Rotation detection in chest radiographs based on generalized line histogram of rib-orientations," in *2014 IEEE 27th International Symposium on Computer-Based Medical Systems (CBMS)*, IEEE, 2014, pp. 138–142.

10

Pathological Medical Image Segmentation: A Quick Review Based on Parametric Techniques

M. Ravi and Ravindra S. Hegadi

CONTENTS

10.1 Introduction

The segmentation procedure is a separation of images into a number of subregions based on a precise characteristic in order to pick up a region of attention. The segmentation process has vast applications in the medical field. A large amount of work has been done to trounce the problems faced by the segmentation procedure, and so far there is a need of further effectual and well-organized work.

10.1.1 Role and Flow of Medical Image Segmentation

The main reason for using medical image segmentation is to resolve the issues in the diagnosis of mankind's deadly diseases. The procedure of segmentation of a medical image might be an optimal solution. This process is extremely dependent on the clinical application of the problem [1–3]. The intention of segmentation is to accelerate the process of visualization to grip the detection process more efficiently. The flow of medical image segmentation is shown in Figure 10.1.

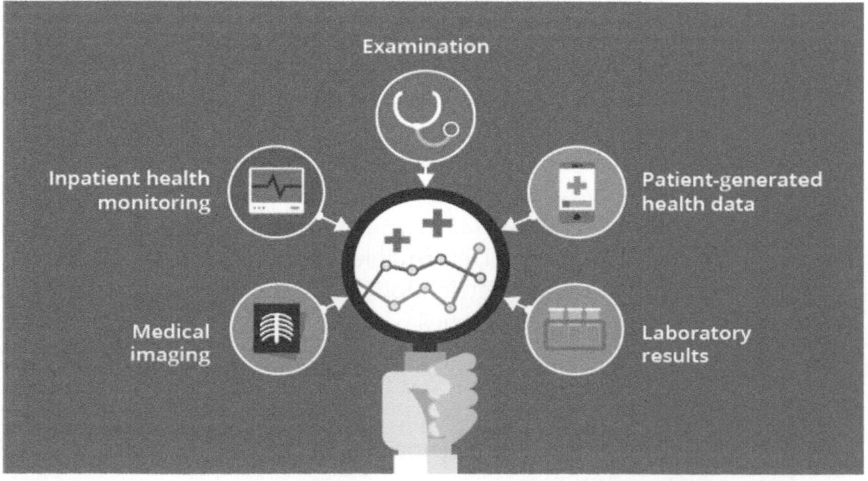

FIGURE 10.1
Role and flow of medical image analysis.

10.1.2 Challenges in Medical Image Modalities for Segmentation

Segmentation of medical images faces many challenges that affect the quality of this process [4]. These problems can be observed in Figure 10.2. The problem of ambiguity arises when there is noise in the image, which makes the classification of an image difficult [5]. The pixels value intensity is amended due to noise in the image. This alteration in the intensity values of the pixels disturbs uniformity in the intensity range of the image [6]. Motion in the picture causes noise, blurring effects, lack of diverse features, etc. The problem of partial volume averaging causes the issue of inconsistency in the intensity values of the image pixels. To manage this ambiguity, image segmentation plays a vital role in medical diagnosis systems [7]. The relevant terminologies used Figure 10.2 are expressed in Table 10.1.

10.1.3 Architecture of Medical Image Modalities

Many imaging techniques have been developed and are in clinical use. Because they are based on different physical principles [11,12], these techniques can be more or less suited to a particular organ or pathology. In practice, they are complementary, because they offer different points of view for the same medical problems. In medical imaging, these different imaging techniques are called modalities. Anatomical modalities provide insight into the anatomical morphology. They include radiography, ultrasound (US), computed tomography (CT), and magnetic resonance imagery (MRI). There are several derived modalities that sometimes appear under a different name, such as magnetic resonance angiography (MRA, from MRI), digital subtraction angiography (DSA, from X-rays), computed tomography angiography (CTA, from CT), etc. Functional modalities, on the other hand, depict the metabolism of the underlying tissues or organs. This list is not exhaustive, as new techniques are being added every few years [13]. Almost all images are now acquired digitally and integrated in a computerized picture archiving and communication system (PACS) shown in Figure 10.3.

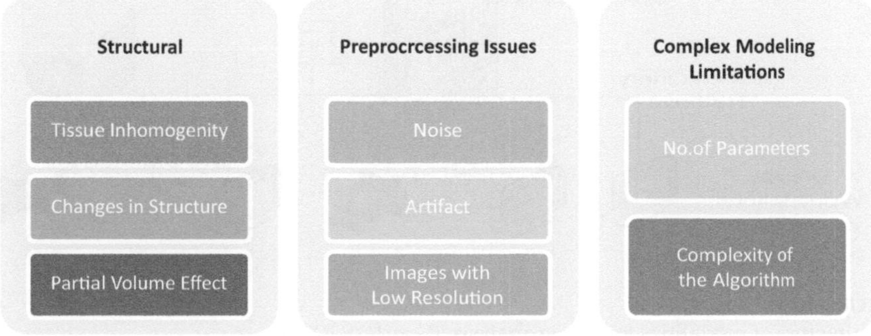

FIGURE 10.2
Challenges in segmentation.

TABLE 10.1

Medical Imaging Terminologies

Terminologies Mentioned in the Figure above	
Tissue Inhomogeneity	Abnormal tissues with non-homogeneous structure.
Changes in Structure	Changes in structure variation in the physical appearance of the image.
Partial Volume Effect	It can be defined as the loss of apparent activity in small objects or regions because of the limited resolution of the imaging system.
Noise	Image noise is a random variation of brightness or color information in images.
Artifact	In natural science and signal processing, an artifact is any error in the perception or representation of any information, introduced by the involved equipment or modality.
Images with Low Resolution	Image resolution is the detail an image holds. The term applies to raster digital images, film images, and other types of images with low value.
Number of Parameters Used	Two parameters: APD and DICE.
Complexity of the Algorithm	Time complexity is a concept in computer science that deals with the quantification of the amount of time taken by a set of code or algorithm to process or run as a function of the amount of input.

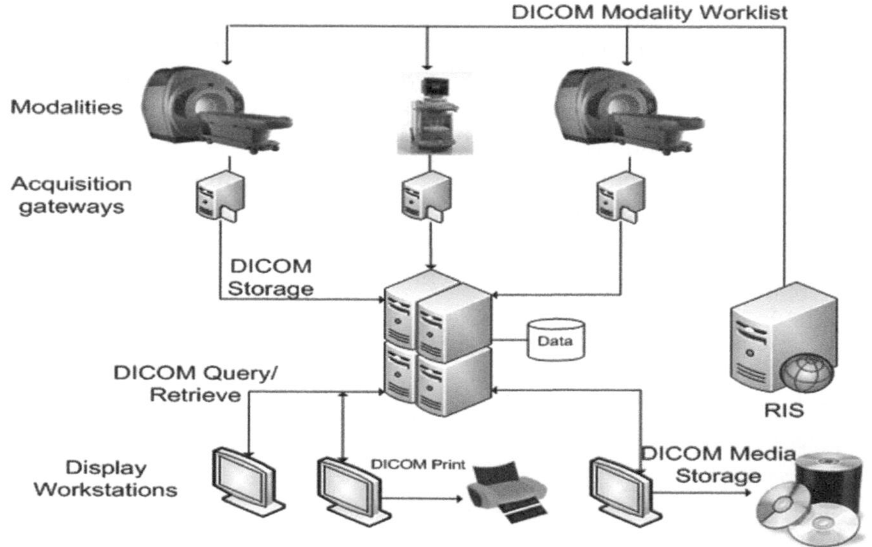

FIGURE 10.3
Architecture of modalities for image analysis. (Courtesy PACS.)

10.1.3.1 MRI

Analyzing these applications in the medical field of segmentation we can say that the majority of work is done using MRI for brain images (see Figure 10.4). Enhancement of the noisy images is further required for the accurate segmentation of the region of interest. Another issue regarding these images is that they contain a variety of resolution, due to which segmenting the image with the required level of contrast is a great problem. The main applications in this regard are extracting the volume of the brain, segmenting different issues in a matter of gray, white cerebrospinal liquid, and outlining precise brain formations [14,15].

10.1.3.2 Electron Microscopy

Electron microscopy (EM) is a technique used in the field of pathology to obtain biological and non-biological specimen images with high resolution. It is used in biomedical research to examine detailed macromolecular complexes and the structure of tissues, cells, and organelles. Microscopic images outcome obtained with high resolution by using electrons (which have very short wavelengths) as the source of enlightening radiation is shown in Figure 10.5. EM is used in combination with a range of ancillary techniques such as thin sectioning, immune-labeling, and negative staining to answer specific questions. EM images provide key information on the structural basis of cell function and of cell disease. Generally, it can be stated that EM is of high value in the investigation of clinical specimens related to renal diseases, tumor processes, and storage disorders [8, 9, 10].

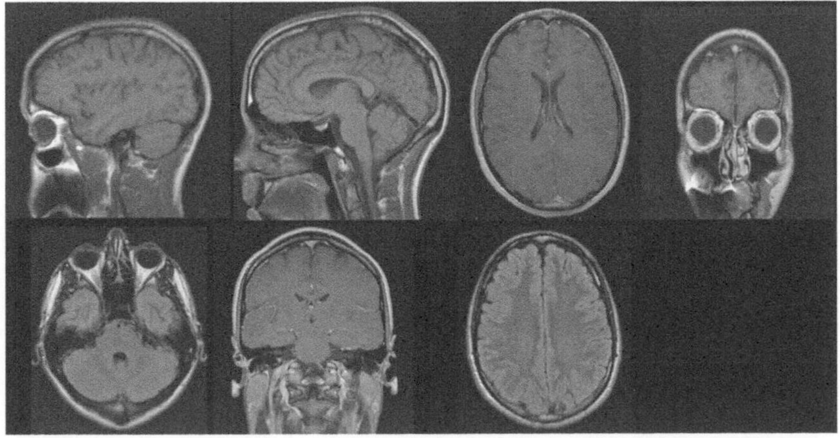

FIGURE 10.4
MRI modality images.

10.1.3.3 Computed Tomography

The analysis of CT images uses the segmentation process with many applications. The major use of the segmentation process in this aspect is in the analysis of bones; thoracic scans; stomach, brain, and liver images; segmentation of heart; and demarcation of abdominal aortic aneurysms [16]. The contrast and resolution of these images are not as good as in MRI images. A variety of methods is applicable in the segmentation process of CT images, as shown in Figure 10.6.

10.1.3.4 US

Usually, US images have elevated rates of defects, which makes it difficult to segment out the region of interest accurately. This makes the segmentation of US images challenging. In spite of this issue, some work has been done in this

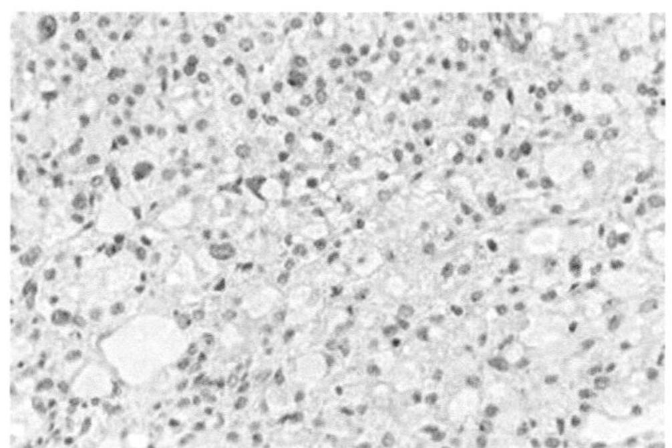

FIGURE 10.5
Electronic microscope modality images.

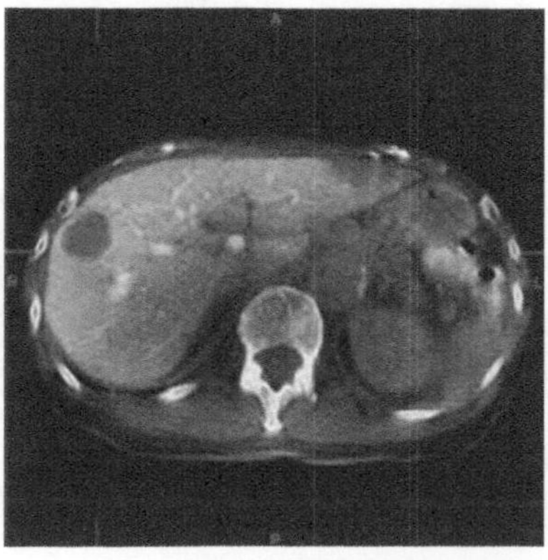

FIGURE 10.6
CT modality images.

regard. In most cases, manual segmentation is done, but these images are also used for the estimation of the motion involved together with the identification of a pathology by means of textural classifiers [17,18] (Figure 10.7).

10.2 Medical Image Segmentation Techniques

Through the history of methods and techniques in the context of medical image segmentation, we find that there is a great improvement in this regard. With the passage of time, more effective and efficient processes have been discovered, as shown in Figure 10.8. Here in this section, we will analyze different methods that have been developed and utilized in the process of medical image segmentation considering recent work done in this field.

10.2.1 Thresholding

Thresholding is a standout process amongst the most widely recognized techniques utilized for picture division. The reason is that it is the best way when we need to dissect the forefront setting by dispensing with the picture foundation. The essential work of this strategy is reliant on the power estimations

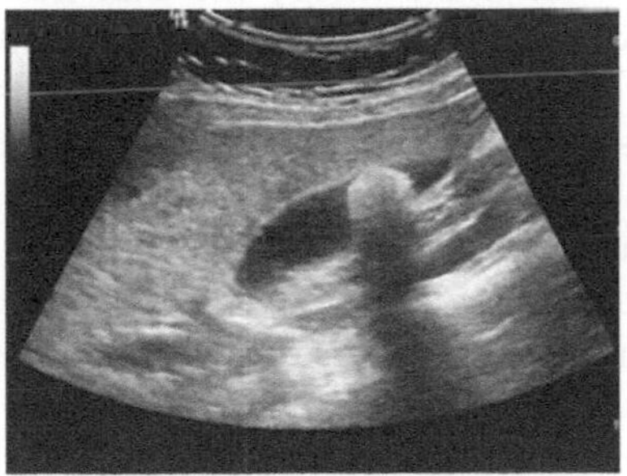

FIGURE 10.7
US modality images.

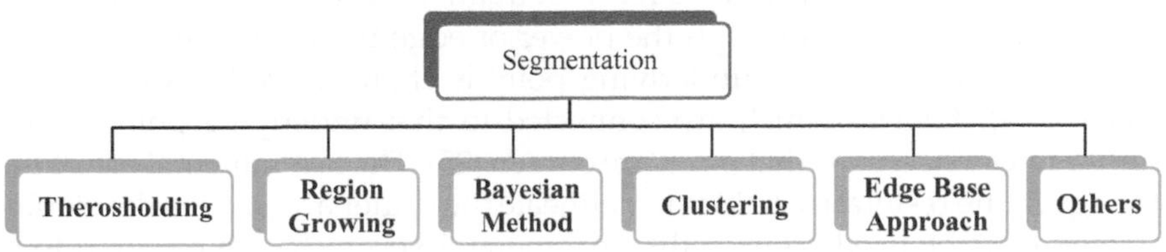

FIGURE 10.8
Categories of medical image segmentation.

of pixels in the picture. The front and background division of the image is done by edge estimation. Extra tasks are expected to wipe out the commotion factor from the picture and to secure more powerful outcomes during the time spent for division [19]. For this situation, a picture is first changed over into a parallel picture and afterwards a characterized edge esteem is utilized which isolates the distinctive areas of the picture. Some ongoing work that has been done is this prospect can be seen in Li et al. [20]. The examination is essentially an outline of picture division systems in radiance of thresholding process. The five strategies of thresholding discussed in this paper incorporate mean strategy, p-tile strategy, histogram dependent technique (HDT), edge maximization technique (EMT), and visual procedure. Another thresholding methodology of therapeutic picture division is exhibited in Ng et al. [21]. This approach uses watershed division together with surface-based area consolidating strategy. The outcomes obtained through this technique are 92.2% when contrasted with the past manual division approaches. SVM thresholding of restorative pictures is portrayed in Kotropoulos and Pitas [22]. This technique is proposed for wavering recognition from a chest X-beam picture. Results acquired through this procedure are attainable. The work is completed by first sifting the picture and fragmenting through limit calculation. After that, a portion of the work is set in SVM to section out the picture unequivocally. Delicate thresholding of medical pictures is proposed in Khare and Tiwary [23]. The strategy takes a shot at the idea of enrollment work that groups each picture pixel to an alternate environment through the picture histogram. The benefits of this strategy are that the technique works consequently and does not require human association, and all the hard choices are put off to the conclusive stage for the examination of spatial tasks. Spatial activities make this strategy stronger and more successful. Mammogram division is depicted in Cascio et al. [24]. An outline of thresholding techniques is introduced, and a step by step limit checking strategy is utilized to settle a normal edge as an incentive for the image. The technique demonstrates that by disposing of sharpness of edges inside a picture, fantastic consequences of division can be observed.

10.2.2 Region Growing

In this procedure, a region of the plot is received through a predefined condition. The condition characterized for this situation depends on the data expert outcome accomplished through the power or edge points of interest of a picture. In this technique, an underlying point is characterized physically, and after that all focuses which are connected to that underlying point having similar force esteems as that point are chosen [25]. The fundamental use of this strategy in the restorative field is to delineate the tumor areas. Area developing strategy can't be used without this technique. Extra tasks are required to be performed before the use of this strategy. The principal challenge of this technique is that it requires manual delineation of the underlying point, in view of which there is a need to state an underlying point for each area that will be extricated. Presently, a concise examination of some ongoing area developing strategies in the restorative picture division prospect will be exhibited. A new method for brain abnormalities segmentation is proposed in Siddique et al. [26]. The method works on the concept of seed-based region growing. It takes the adult male and female MR images of different sizes from a brain. Brain tissues and background are divided into different categories and receive different-sized MR images as input. The results are promising, and light images yield better results as compared to darker abnormalities images. A similar approach is given in Oghli et al. [27]. A method is an automatic approach of region growing. A co-occurrence matrix is used that selects a starting point for the seed. The advantage acquired from the method is that it reduces the time factor for the manual post-processing process. MRI brain tumor segmentation is described in Poonguzhali and Ravindran [28]. The method focuses on gradients and variations along with the boundaries. Edge information is preserved through the use of the anisotropic filter. Next, the mean variance and mean gradient of the boundary curve are calculated.

This method is effective for categorizing affected regions. Another similar US segmentation method is proposed in Poonguzhali and Ravindran [29]. A method is an automatic approach for masses segmentation from US images. The method can be said to be optimal for the segmentation of US images because they preserve the spatial information and are insensitive to speckle noise. A hybrid approach to US image segmentation is presented in Guan et al. [30]. This method is a composition of two approaches: region growing and region merging. Effective results are achieved through this. A Bayes-based medical image segmentation approach is proposed in Pan and Lu [31–38]. This method works by adjusting parameters during the region growing approach. It is a multi-stage processing approach that yields effective results as it is insensitive to noise and decreases the computational time to a great extent.

10.2.3 Bayesian Approach

Bayesian assessment theory is used for classification purposes. The method mainly works by considering probability in the image to construct models based on the probability that is further utilized for the class assignment of voxels in the image. These voxels are treated as random variables in the image. There are four main approaches in the Bayesian category of image segmentation. These approaches are shown in Figure 10.9.

10.2.3.1 Maximum a Posteriori (MAP)

In Bayesian figures, MAP approximation is an approach of the posterior allotment. MAP can be employed to acquire a summit approximation of an unnoticed measure taking place in the foundation of experimental information.

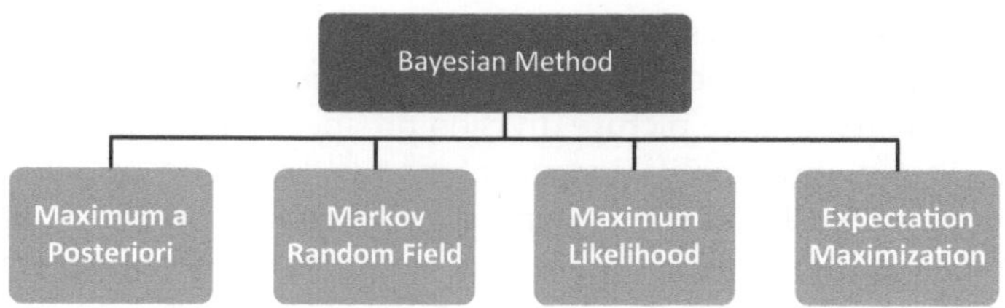

FIGURE 10.9
Bayesian approaches.

It is intimately interrelated to Fisher's technique of maximum likelihood (ML) method, although it occupies an amplified optimization purpose which integrates a preceding allotment above the measure one desires to approximate. MAP evaluation can be observed as a regularization of ML evaluation [39].

10.2.3.2 Markov Random Field (MRF)

MRF basically makes use of the undirected graph that determines the Markov values of some arbitrary variables contained within a graphical model. A MRF is quite similar to the Bayesian approach in view of representation. The only difference is that this approach is undirected whereas the Bayesian method is comprised of directed graphs [40].

10.2.3.3 Maximum Likelihood (ML)

This is a phenomenon in statistics that aims at providing the estimation of parameters in a given statistical model. Furthermore, it is regarded as a well-recognized estimation technique. In some scenarios, it is also used to maximize the likelihood function when we are given a fixed amount of data together with its statistical model from where values are selected for the parameters that carry out an overall job of maximization [41].

10.2.3.4 Expectation Maximization (EM)

This is also a statistical approach used to figure out the MAP or ML of parameters of a statistical model. This approach works on the basis of iterations. Here, steps are performed in alterations; first, an estimation (E) step is performed, followed by a maximization (M) step whose information is then utilized for the next E step, and the process goes on [42].

10.2.4 Clustering

If we compare the functions of clustering and classifiers, we can say that both are carrying out the same function with a difference in their way of working. Classifiers make use of training data to classify the image and thus are

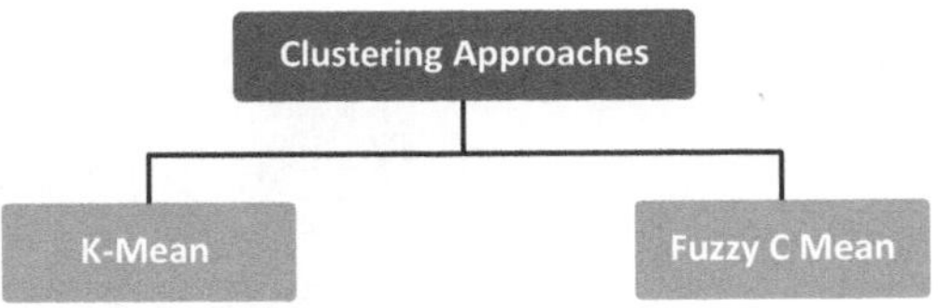

FIGURE 10.10
Clustering approaches.

called supervised methods. A clustering approach contains unsupervised methods as it does not make use of training data. This inability of learning in the clustering approach is compensated by iteratively dividing the image through the segmentation process and then illustrating the possessions of every division. In other words, we can say that clustering techniques instruct themselves by means of existing statistics [43]. The clustering method is useful for the applications where there is a presence of disjoint sets of various pixel intensities in the image. The main application of this method can be observed in the segmentation of MRI. There are two main methods of clustering approach which are commonly used for the segmentation of medical images. These methods are shown in Figure 10.10.

10.2.4.1 k-Means

The clustering process, in this case, is carried out by iteratively calculating the mean of intensities values of each separated class or cluster of the image. And the segmentation is carried out by categorizing each pixel with the closest obtained mean of the image [44]. This method is evaluated on the affected area of the renal cell carcinoma (RCC) image. We detected and segmented the area affected by RCC. Figure 10.11 shows some images of the area affected by RCC segmented using the proposed method [96].

10.2.4.2 Fuzzy C-Means

Segmentation through this process is carried out on the basis of a fuzzy set premise. This process is also called generalization of a k-means process. The difference between the two processes is that the points are categorized in separate classes in the k-means process, whereas fuzzy c-means allows the points to be connected with more than one class [45]. Now we will have a brief review of some recent work done in this regard. Segmentation of MR images using the fuzzy c-means approach is presented in Li et al. [46]. A new version of the fuzzy method is proposed that makes it possible to determine automatically the required number of clusters for the segmentation process. This method makes use of statistical histogram information to achieve its task. The results showed that the method produces more accurate results as compared to the classic fuzzy means approach. Another similar approach

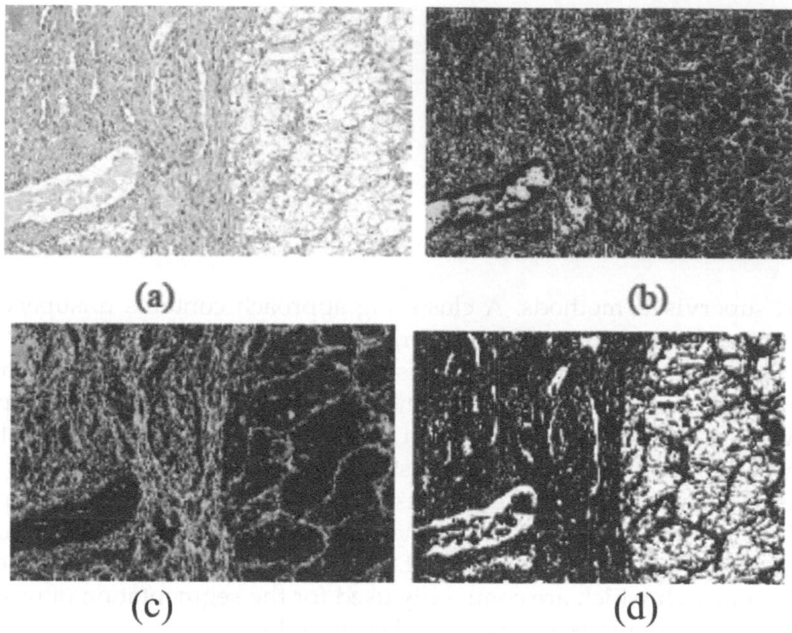

FIGURE 10.11
K-means clustering for infected with RCC disease (a) original image, (b) first cluster, (c) second cluster, (d) third cluster image is the final output where the tumor area i.e. the affected area is clearly segmented.

for MRI segmentation is proposed in Ozyurt et al. [47]. In order to reduce the computational time for the segmentation process, a new method is proposed in Szilagyi et al. [48]. This method works based on the BCFCM method by introducing a new factor. The results show that it is a quick and optimal approach for the endoscopy of the brain. The fuzzy means approach together with the dominated grey levels in the image can be observed in Balafar et al. [49]. The method works by converting the image into gray level and reducing the noise by applying the wavelet approach. Then the image is clustered by taking into consideration central gray levels as a base for the clustering process. A similar approach is presented in Birgani et al. [50]. MRI segmentation using fuzzy c-means and neural networks is presented in Saripan et al. [51]. The results acquired through this combination are robust, fast, and accurate even in the presence of high noise. An approach to improve the speed of the segmentation process based on k-means is presented in Nandagopalan et al. [52]. The results showed that this approach is very fast and reliable and provides good quality as well. Volume-based medical image segmentation using the k-means approach is proposed in Li [53]. The process works by initially preprocessing the image to speed up rest of the processes. Then, different clustering approaches are analyzed and a new approach is presented to accurately segment out images and to speed up the segmentation process. Using the trained k-means clustering approach for MRI segmentation is described in Kumbhar and Kulkarni [54]. The k-means approach is used to segment white and gray matter from the MR images. High precision is achieved through this method as compared to the classic k-means approach. A new approach for the segmentation of CT images is presented in Rathnayaka et al. [55]. The method works in three sections. The sections are grouped on the basis of detecting abnormal regions, cerebrospinal fluid, and lastly brain matters. The results acquired through the process are not outstanding. The most recent work in this regard can be seen in Sinha and Ramakrishnan [56]. The method is basically a blood cell segmentation approach using the k-means and median cut approach. Initially, the best outcome of blood cell segmentation is analyzed through the k-means, fuzzy c-means, and mean shift approaches, and then the median cut approach is applied. The results achieved through the process showed that it is better to process for object segmentation when further feature extraction process is needed. A mountain-based clustering approach for medical image segmentation is presented in Ng et al. [57]. The results achieved are effective for diagnosing many issues in the medical field. A combination of clustering approaches involving fuzzy means, Bayesian method, and user interaction is proposed in Saripan et al. [51]. Medical image segmentation using k-means together with the watershed approach is presented in Ng et al. [57]. Medical image segmentation using fuzzy similarity relation is presented in Tabakov et al. [58].

10.2.5 Deformable Methods

The deformable method works on the basis of object boundaries. The features considered in view of image boundaries are shape, smoothness, and internal forces together with the external forces on the object under consideration [59]. All these factors influence the effectiveness of the obtainable results. Closed curves and shapes in the image are utilized to outline the object boundaries. The process of outlining the boundary of an object is a closed curvature or plane that is initially positioned close to the preferred edge and later permitted to experience an iterative reduction progression. In order to maintain the segmentation process, smooth internal forces are derived from the image. The external forces are derived in order to originate a plane towards the preferred element in the image. The main advantage of these methods is the piecewise continuity. Deformable forces are mainly classified into two different categories, as shown in Figure 10.12.

10.2.5.1 Parametric Deformable Models (Explicit)

In the statistics of deformable models, parametric models are the ones that can be described using a finite number of parameters. These methods are also called active contours and make use of parameter-generated curves for the representation of shape models.

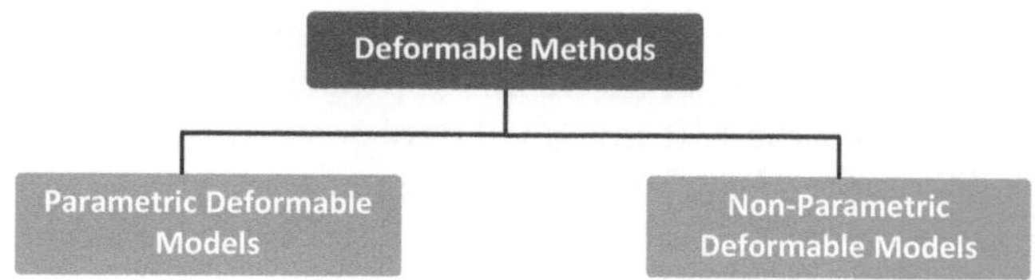

FIGURE 10.12
Deformable methods classification.

Parametric models are further divided into two categories, which are:

- Edge-based methods
- Region-based methods

The methods in the edge-based category take edges information as image features for the segmentation process and are really responsive to the noise factor, as any noise can alter the accurate information of edges [60]. The other category under parametric methods is that of region-based methods, which make use of different areas to segment out the image. In the model evaluation process, the information of regions is not updated, which makes it difficult to obtain any changes in the region features. The main drawback of these methods is that it is difficult to handle the topology changes in the anonymous entity segmentation [61].

10.2.5.2 Non-Parametric Models (Implicit)

These methods are also called geometric active contour methods. They are level set approaches and are based on the concepts of convolution theory. In the process of defining the curve for the segmentation, a level set function is utilized together with the additional time aspect [62]. The evaluation of the curve in this case is independent of parametric values. The drawback of parametric models is handled through these models, which allow automatic handling of variations in the topological factors. These methods also make use of edge-based and region-based methods, although their implementation process is different from that of parametric models [63]. Now we will have a short overview of some of the most recent techniques proposed in this regard. In Tsechpenakis and Metaxas [64] we can analyze 3D medical image segmentation through the combination of conditional random fields and deformable models. The results acquired through this approach are promising when compared to previously developed methods. Medical image segmentation using minimal path deformable model is presented in Yan and Kassim (2006) [65]. This approach is based on extracting the organ contours. It was considered as one of the great achievements in the segmentation of various types of medical images. Similar work can be seen in Yan and Kassim (2004) [66]. A geometric deformable model for medical image segmentation is proposed in Lee et al. [67]. Medical image segmentation using a combination of genetic algorithm and non-convex approach is presented in McIntosh andHamarneh [68]. The typical gradient of deformable model in this case is replaced with the genetic algorithm, as it is assumed that the genetic algorithm cannot provide optimal solutions, but when it is combined with deformable models satisfactory results are achieved. The implicit-shaped deformable model for medical image segmentation is presented in Farzinfar et al. [69]. This method makes use of region-based and statistical model-based approaches to extract an object. The use of shape and appearance priors in the deformable models for medical image segmentation is presented in El-Baz and Gimel'farb [70]. The main advantage acquired through this method is fast processing speed. A fuzzy non-parametric approach for medical image segmentation can be observed in Awate et al. [71]. This method provides more accuracy as compared to the tractography approach in this prospect. Medical image segmentation using a local binary fitting approach and deformable models is presented in Nakhjavanlo et al. [72]. It has the main advantage of speeding up the curve evaluation process. A survey on deformable models is given in Sonka and Fitzpatrick [73]. Similar work is also presented in Xu et al. [60]. The non-parametric mixture model–based medical image segmentation is described in Joshi and Bradi [74].

10.2.6 Atlas-Guided Approaches

Medical images segmentation based on atlas-guided approaches is a way of analyzing image through labeling a preferred structure or set of framework commencing images made through modalities of medical imaging. The main purpose of this approach is to lend a hand to radiologists in the discovery and identification of diseases. The working flow of this approach is optimized by identifying significant anatomy in the medical images [75]. These approaches are also called adaptable templates. The segmentation, in this case, is carried out by preparing an atlas using compiled information of anatomy. After the generation of the atlas, this is used as a reference structure for the segmentation of fresh images. These approaches consider the registration problem to handle the segmentation process. Atlas wrapping is used for the segmentation process which works by mapping the generated atlas on the objected image [11]. The main application of these approaches is in the images where there is no well-defined relationship between image pixels and regions. The other main applications include their use in clinical practice and computer-aided diagnosis to analyze the shape and morphological differences between image regions.

10.2.6.1 Atlas as Average Shape

In this category, we have two approaches, which are described in the subsections below.

10.2.6.1.1 Active Shape Model (ASM)

The ASM makes use of principal component analysis (PCA) and a pre-constructed shape model for the segmentation of medical images. All shapes are trained and aligned and are used with PCA for the purpose of segmentation [76]. The work is carried out by using the average mean shape for the scan and then deformation is carried out by means of deformable models. The process is performed through iterations and in each iteration, the previous utilized curve or shape is used to measure the desired target object. The parameters of shape also remain unchanged. By means of this process, only the desired deformations are permitted and the process terminates when variations are faced in the shape model [77].

10.2.6.1.2 Active Appearance Model (AAM)

The AAM works in the same way as ASM with the difference that together with the shape model it also makes use of the intensity model for the purpose of segmentation. The intensity model is generated through registration among the training statistics [78].

10.2.6.2 Atlas as Individual Image

The process works by manual segmentation of anatomical structures originated from a reference image which creates a spatial atlas or map. The registration of the reference image is done on the atlas for the purpose of automatic segmentation. The mapping among the two should be coordinated. After that, intensity correspondence evaluation is carried out using different approaches. The most used methods, in this case, are cross-correlation and mutual information for registration. The smoothness factor is handled through the Gaussian or elastic models [79]. Now we will have a brief overview of some recent approaches in this regard. The latest work in this field can be analyzed in Chen et al. [80]. The approach works on the basis of the graph cut method by combining it with the AAM. The three main steps of this approach include model building, object recognition and, lastly, delineation. The method was tested for the segmentation of liver, kidney, and spleen images. Overall, the obtained results show an accuracy of 94.3%. Segmentation of spinal images through AAM can be analyzed in Chen et al. [81]. A different work carried out in this approach makes use of a combination of ICA and AAM to segment out the spinal images. The results showed that this approach contains more accuracy than the traditional PCAAAM approach. The part of the ASM in this prospect can be analyzed in Chen et al. [80]. Medical image segmentation using the statistical shape model is presented in Neumann and Lorenz [82]. This approach works by combining the point distribution fraction with two dimensional PCAs to segment out medical images.

The results showed improvement as compared to the traditional approach. Segmentation of pelvic X-ray images through splines and shape modes is proposed in Safavian and Landgrebe [83]. The deformation process is enhanced in this case by combining the ASM with the cubic spline interpolation approach due to low resolution and blurring effects within the X-ray images. The results showed improvement even in the presence of a fracture. A 3D ASM for MRI image segmentation is carried out in Rousson et al. [84]. Another 3D segmentation of MRI using the level set approach is presented in Baillard et al. [85]. This method includes 3D filtering followed by 3D segmentation of MR images. Graph cut based DT-MRI segmentation can be analyzed in Weldeselassie and Hamarneh [86]. Similar work is carried out in Stawiaski et al. [87]. Segmentation of liver tumors using graph cuts and a watershed approach is presented in Peng et al. [88]. Atlas registration–based cerebellum MRI segmentation can be analyzed in van der Lijn er al. [89].

10.2.7 Edge-Based Approaches

These approaches are the most common way of detecting discontinuity and boundaries of objects within an image. This method is the most common way of detecting pixels with the same intensity level of an object. In this case, the two connected pixels have same intensity distribution form the edge and it is not essential that they form a closed path [90]. The distinction between the pixels, in this case, is carried out by estimating the intensity gradient. These methods are mainly used as a base or central technique for other segmentation approaches [91].

10.2.8 Compression-Based Approaches

Compression-based techniques assume that the best possible segmentation is the one that reduces the excess of every achievable segmentation and the development period of statistics. The association connecting these two conceptions is that segmentation attempts to discover samples in an image and reliability in the image might be utilized to compress it. This technique explains every division by means of its surface and edge outline [92]. On behalf of any specified segmentation of an image, this method gives the number of bits essential to predetermine that the image is centered on known segmentation. Consequently, surrounded by any achievable segmentation of an image, the aim is to discover segmentation that generates the straight coding span. This might be accomplished through an easy clustering technique. The deformation inside the lossy compression decides the roughness of segmentation and its best assessment could be different for every image. This limitation can be projected, commencing the disparity of consistency in an image. For instance, when the textures in an image are alike, such as in disguise images, stronger compassion and minor quantization are necessary.

10.2.9 Other Techniques

There are numerous other techniques proposed and developed in view of medical image segmentation. Some of these methods include a watershed algorithm, model fitting, partial differential equation-based methods, split-and-merge methods, fast marching methods, and multi-scale segmentation. Some of the techniques presented in this regard can be analyzed here. Medical image segmentation using the watershed transform is presented in Jia-xin and Sen [93]. This method works on the basis of morphological operations to extract objects from medical images. An improved watershed transforming process can be analyzed in Li, Wu, and Sun [94]. Medical image segmentation using feature-based GVF snake is presented in Harini and Chandrasekar [95]. The Chan-Vese display for dynamic shapes, which is an adaptable and effective strategy, is ready to fragment different classifications of pictures, and, in addition, some that would be to some degree entangled to portion utilizing of conventional division, for example, inclination-based and thresholding strategies. With a variety in the parameter estimations of the Chan-Vese [97] display, better yield acquired in segmented images is mentioned in Figure 10.13.

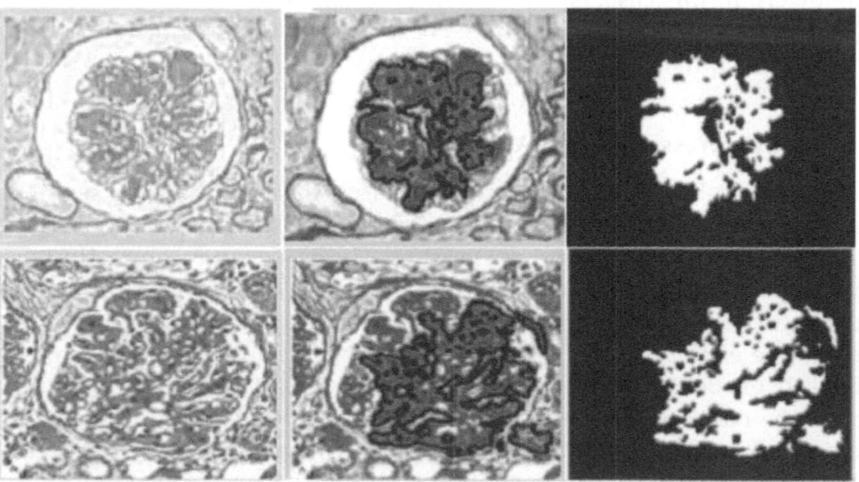

FIGURE 10.13
Counter-based segmentation of glomerular images.

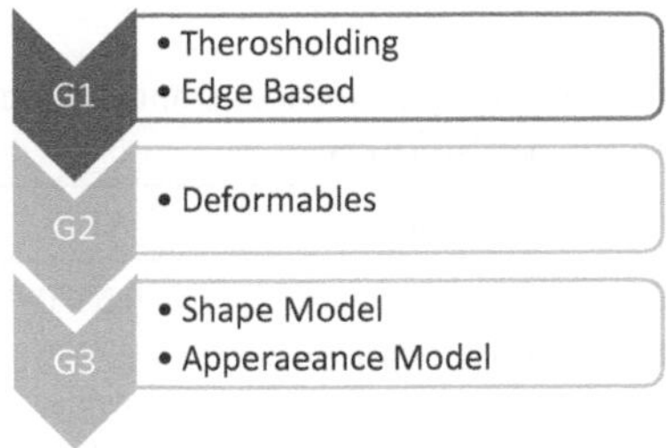

FIGURE 10.14
Various generations of medical segmentation approaches.

10.3 Study and Conversation

After seeing an overview of basic approaches in the medical image segmentation process, it can be noticed that, with advancements in the process of search and development, new effective and efficient approaches are coming into existence. The methods classification according to generation can be analyzed to understand this point. Approaches that exist in this regard are categorized into three generations: first, second, and third respectively, as shown in Figure 10.14.

Development and progress in the field of medical image segmentation can be analyzed by placing the methods into different groups. This classification will show us the research and development carried out in this regard. The first generation group contains methods that require little prior information for processing the image and hence involve low-level techniques. With the passage of time and advancement in technology, some new and more effective methods came into existence. The second generation includes methods based on optimization, image, and uncertainty models. Third generation techniques are highly dependent on prior information of the image and require experts' defined models and rules for the classification of an image. A discussion can also be started on the comparison of the different methods given in Table 10.2. This will help us to select the best method for a given situation.

10.4 Comparison of Medical Image Segmentation Methods with Experimental Analysis

We have conducted experiments over different segmentation techniques. We have used Matlab-2016a for our experimentation and a total of 117 images. It was found that the region-based, edge-based, and k-means

TABLE 10.2

Experimental Results of Medical Image Segmentation Using Different Techniques

Sl. No	Methods	APD	DICE
1	**Threshold**	0.5370	0.9662
2	**Adaptive Thresholding**	4.7835	0.7811
3	**Edge Based**	0.7539	0.9421
4	**Region Based**	0.7835	0.9454
5	**Level Set**	5.2800	0.7514
6	**K-Mean**	0.8680	0.9345
7	**Fuzzy C-Means**	0.6330	0.9604

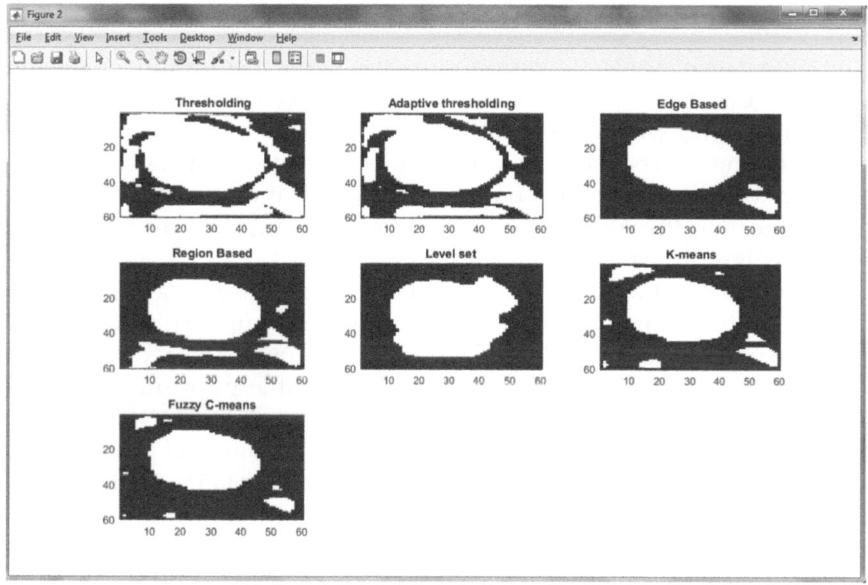

FIGURE 10.15
Resultant images segmented with various segmentation techniques.

techniques are efficient and reliable with respect to computation time and accuracy of segmentation. We have used two parameters, namely asymmetric partition distance (APD) and DICE in order to compare different image segmentation techniques, as shown in Figure 10.15 and in Table 10.2.

A graphical representation of the above-mentioned methods can be found in Table 10.3 and Table 10.4.

TABLE 10.3

Graphical Representation of the Different Medical Image Segmentation Technique

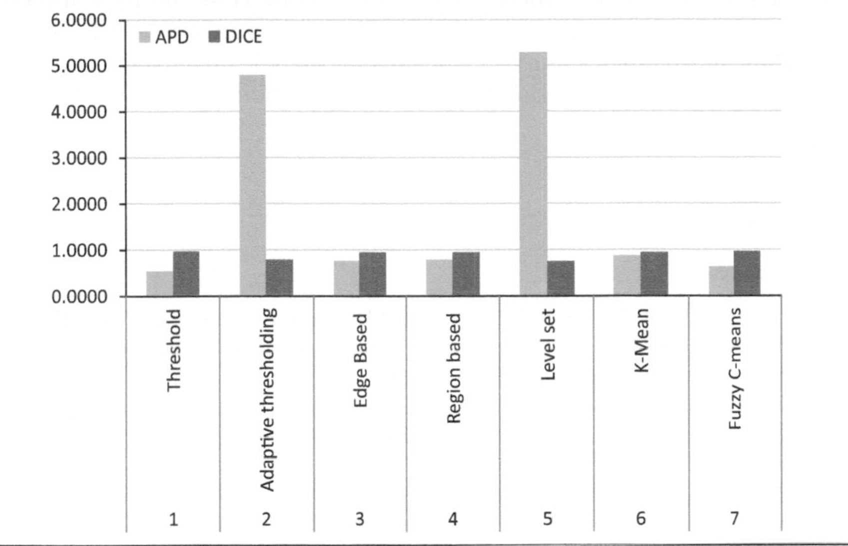

TABLE 10.4

Summary of Comparison Using Different Medical Image Segmentation Techniques

Technique	Thresholding	Region-Based	Edge-Based	Clustering
Benefits	Simple	Specific for Segmentation	Specific for Computational Factor	Easy to Implement
Restrictions	Responsive for Artifacts and Noise	Computationally Complex	Not Appropriate for all Images	Spatial Constraint
Field of Application	Structural intensity	Effective on CT and MRI Images	Suitable for All Medical Modalities	Suitable for All Medical Images
Speed to accomplish the task	Fastest	Slow	Fast	Medium

10.5 Conclusions

This paper provides a brief overview of some methods and techniques available under the umbrella of medical image segmentation. The medical field is comprised of a number of medical modalities and each one of them contains a number of diseases and issues under its heading. Thus, this paper is basically analyzing the techniques proposed. Each method has its own pros and cons. The usage of each method depends on the type of application built together with the resources available. However much research work has been done in this regard we can still say that there is huge room available for finding optimal segmentation techniques for pathological medical imaging.

References

1. T. Zueva, S. O. Oludayo OOjo, and S. M. Ngwira, "Image segmentation available techniques, developments and open issues," *Canadian Journal on Image Processing and Computer Vision*, vol. 2, no. 3, p. 209, 2011.
2. N. Sharma and L. M. Aggarwal, "Automated medical image segmentation techniques," *Journal of Medical Association of Medical Physicists of India*, vol. 35, no. 1, p. 3, 2010.
3. H. Costin, "A fuzzy rules-based segmentation method for medical images analysis," *International Journal of Computers, Communications & Control*, vol. 8, no. 2, 2013.
4. V. Shrimali, R. Anand, and V. Kumar, "Current trends in the segmentation of medical ultrasound b-mode images: a review," *IETE Technical Review*, vol. 26, no. 1, p. 8, 2009.
5. W. Birkfellner, *Applied Medical Image Processing: a Basic Course*, Taylor & Francis, 2014.
6. S. S. Al-Amri, N. V. Kalyankar, and S. D. Khamitkar, "A comparative study of removal noise from remote sensing image," *International Journal of Computer Science Issues (IJCSI)*, vol. 7, no. 1, 2010.
7. B. S. He, F. Zhu, and Y. G. Shi, "Medical image segmentation," *Advanced Materials Research*, vol. 760, pp. 1590–1593, 2013.
8. H. U. Zollinger, M. J. Mihatsch, "Renal pathology in biopsy: light, electron and immunofluorescent microscopy and clinical aspects," Springer Science & Business Media, 2012.
9. Y. Collan, P. Hirsimäki, H. Aho, M. Wuorela, J. Sundström, R. Tertti, K. Metsärinne, "Value of electron microscopy in kidney biopsy diagnosis," *Ultrastructural Pathology*, vol. 29, no. 6, pp. 461–468, 2005.
10. M. A. Zariwala, M. R. Knowles, H. Omran, "Genetic defects in ciliary structure and function," *Annual Review of Physiology*, vol. 69, pp. 423–450, 2007.
11. Rumen Mironov and Roumen Kountchev, "Architecture for medical image processing," *Advances in Intelligent Analysis of Medical Data and Decision Support Systems Data and Decision Support Systems*, 2013, pp. 225–234.
12. M. Sharif, M. Yasmin, S. Masood, M. Raza, and S. Mohsin, "Brain image reconstruction: a short survey," *World Applied Sciences Journal*, vol. 19, no. 1, pp. 52–62, 2012.
13. S. Mohsin, M. Yasmin, M. Sharif, M. Raza, and S. Masood, "Brain image analysis: a survey," *World Applied Sciences Journal*, vol. 19, no. 10, pp. 1484–1494, 2012.

14. M. Sharif, M. Yasmin, S. Masood, M. Raza, and S. Mohsin, "Brain image enhancement – a survey," *World Applied Sciences Journal*, vol. 17, no. 9, pp. 1192–1204, 2012.
15. M. Sharif, S. Masood, M. Yasmin, M. Raza, and S. Mohsin, "Brain image compression: a brief survey," *Research Journal of Applied Sciences*, 2013.
16. C. Xiao-Juan and L. Dan (eds.), "Medical image segmentation based on threshold SVM," in *2010 International Conference on Biomedical Engineering and Computer Science (ICBECS)*, IEEE, 2010.
17. S. Aja-Fernandez, G. Vegas-Sanchez-Ferrero, and M. Fernandez (eds.), "Soft thresholding for medical image segmentation," in *2010 Annual International Conference of the Engineering in Medicine and Biology Society (EMBC)*, IEEE, 2010.
18. M. Sharif, A. Shahzad, M. Raza, and K. Hussain, "Enhanced watershed image processing segmentation," *Journal of Information & Communication Technology*, vol. 2, no. 1, pp. 1–9, 2008.
19. A. Abubakar, R. Qahwaji, M. J. Aqel, and M. H. Saleh (eds.), "Average row thresholding method for mammogram segmentation," in *IEEE-EMBS 2005 27th Annual International Conference of the Engineering in Medicine and Biology Society*, IEEE, 2006.
20. J. Li, S. Zhu, and H. Bin, "Medical image segmentation techniques," *Journal of Biomedical Engineering*, vol. 23, no. 4, pp. 891–894, 2006.
21. H. Ng, S. Huang, S. Ong, K. Foong, P. Goh, and W. Nowinski (eds.), "Medical image segmentation using watershed segmentation with texture-based region merging," *Engineering in Medicine and Biology Society, EMBS 30th Annual International Conference of the IEEE*, IEEE, 2008.
22. C. Kotropoulos and I. Pitas, "22 segmentation of ultrasonic images using support vector machines," *Pattern Recognition Letters*, vol. 24, no. 4, pp. 715–727, 2003.
23. A. Khare and U. S. Tiwary, "Soft-thresholding for de noising of medical images—a multi resolution approach," *International Journal of Wavelets, Multi resolution, and Information Processing*, vol. 3, no. 04, pp. 477–496, 2005.
24. D. Cascio *et al.* "Mammogram segmentation by contour searching and mass lesions classification with a neural network," *IEEE Transactions on Nuclear Science*, vol. 53, no. 5, pp. 2827–2833, 2006.
25. W. Haider, M. Sharif, and M. Raza, "Achieving accuracy in early stage tumor identification systems based on image segmentation and 3D structure analysis," *Computer Engineering and Intelligent Systems*, vol. 2, no. 6, pp. 96–102, 2011.
26. I. Siddique, I. S. Bajwa, M. S. Naveed, and M. A. Choudhary (eds.), "Automatic functional brain MR image segmentation using region growing and seed pixel," in *International Conference on Information& Communications Technology*, 2006.
27. M. G. Oghli, A. Fallahi, and M. Pooyan (eds.), "Automatic region growing method using GSmap and spatial information on ultrasound images," in *2010 18th Iranian Conference on Electrical Engineering (ICEE)*, IEEE, 2010.
28. H. Deng, J. Liu, W. Deng, and W. Xiao, "MRI brain tumor segmentation with region growing method based on the gradients and variances along and inside of the boundary curve," in *3rd International Conference on Biomedical Engineering and Informatics (BMEI)*, IEEE, 2010.
29. S. Poonguzhali and G. Ravindran (eds.), "A complete automatic region growing method for segmentation of masses on ultrasound images," in *2006 ICBPE 2006 International Conference on Biomedical and Pharmaceutical Engineering*, IEEE, 2006.

30. H. Guan, D.-y. Li, J.-l. Lin, and T.-F. Wang (eds.), "Segmentation of ultrasound medical image using a hybrid method," in *2007 CME 2007 IEEE/ICME International Conference on Complex Medical Engineering*, IEEE, 2007.
31. Z. Pan and J. Lu, "A Bayes-based region-growing algorithm for medical image segmentation," *Computing in Science & Engineering*, vol. 9, no. 4, pp. 32–38, 2007.
32. J. C. Bezdek, L. Hall, and L. Clarke, "Review of MR image segmentation techniques using pattern recognition," *Medical Physics*, vol. 20, p. 1033, 1993.
33. L. Le Cam, "Maximum likelihood: an introduction," *International Statistical Review*, vol. 58, no. 2, pp. 153–171, 1990.
34. S. Arya, D. M. Mount, N. S. Netanyahu, R. Silverman, and A. Y. Wu, "An optimal algorithm for approximate nearest neighbor searching fixed dimensions," *Journal of the ACM (JACM)*, vol. 45, no. 6, pp. 891–923, 1998.
35. Y. Liao and V. R. Vemuri, "Use of K-nearest neighbor classifier for intrusion detection," *Computers & Security*, vol. 21, no. 5, pp. 439–448, 2002.
36. D.-Y. Yeung and C. Chow (eds.), "Parzen-window network intrusion detectors," in *2002 Proceedings 16th International Conference on Pattern Recognition*, IEEE, 2002.
37. T. Song, M. M. Jamshidi, R. R. Lee, and M. Huang, "A modified probabilistic neural network for partial volume segmentation in brain MR image," *IEEE Transactions on Neural Networks*, vol. 18, no. 5, pp. 1424–1432, 2007.
38. S. R. Safavian and D. Landgrebe, "A survey of decision tree classifier methodology," *IEEE Transactions on Systems, Man, and Cybernetics*, vol. 21, no. 3, pp. 660–674, 1991.
39. W. M. Wells III, W. E. L. Grimson, R. Kikinis, and F. A. Jolesz, "Adaptive segmentation of MRI data," *IEEE Transactions on Medical Imaging*, vol. 15, no. 4, pp. 429–442, 1996.
40. S. Z. Li and S. Singh, *Markov Random Field Modeling in Image Analysis*, Springer, 2009.
41. G. King, *Unifying Political Methodology: The Likelihood Theory of Statistical Inference*, University of Michigan Press, 1989.
42. M. Yan, A. A. Bui, J. Cong, and L. A. Vese, "General convergent expectation maximization (EM)-type algorithms for image reconstruction," *Inverse Problems & Imaging*, vol. 7, no. 3, 2013.
43. T. Saikumar, K. FasiUddin, B. V. Reddy, and M. A. Uddin, "Image segmentation using variable kernel fuzzy C means (VKFCM) clustering on modified level set method," *Computer Networks & Communications*, pp. 265–273, 2013.
44. T. H. Lee, M. F. A. Fauzi, and R. Komiya (eds.), "Segmentation of CT brain images using K-means and EM clustering," in *2008 CGIV'08 Fifth International Conference on Computer Graphics, Imaging, and Visualisation*, IEEE, 2008.
45. T. Z. T. Muda and R. A. Salam (eds.), "Blood cell image segmentation using hybrid k-means and median-cut algorithms," in *2011 IEEE International Conference on Control System, Computing and Engineering (ICCSCE)*, IEEE, 2011.
46. M. Li, T. Huang, and G. Zhu (eds.) Improved fast fuzzy C-Means algorithm for medical MR images segmentation," in *2008 WGEC'08 Second International Conference on Genetic and Evolutionary Computing*, IEEE, 2008.
47. O. Ozyurt, A. Dincer, and C. Ozturk (eds.), "Brain MR Image segmentation with fuzzy C-means and using additional shape elements," in *2009 BIYOMUT, 2009 14th National Biomedical Engineering Meeting, IEEE*, 2009.
48. L. Szilagyi, Z. Benyo, S. M. Szilágyi, and H. Adam (eds.) "MR brain image segmentation using an enhanced fuzzy c-means algorithm," in *2003 Proceedings of the 25th Annual International Conference of the IEEE Engineering in Medicine and Biology Society*, 2003.

49. M. Balafar, A. R. Ramli, M. Iqbal Saripan, R. Mahmud, and S. Mashohor (eds.), "Medical image segmentation using fuzzy C-mean (FCM) and dominant grey levels of image," in *2008 VIE 2008 5th International Conference on Visual Information Engineering, 2008*; IET, "A survey on medical image segmentation," *Current Medical Imaging Reviews*, vol. 11, no. 1, p. 13, 2015.
50. P. M. Birgani, M. Ashtiyani, and S. Asadi (eds.), "MRI segmentation using fuzzy c-means clustering algorithm basis neural network," in *2008 ICTTA 2008 3rd International Conference on Information and Communication Technologies: From Theory to Applications*, IEEE, 2008.
51. M. Iqbal Saripan, S. Mashohor, M. A. Balafar, and A. R. Ramli, "Medical image segmentation using Fuzzy C-mean (FCM)," in *International Conference on ICWAPR'08*, IEEE, 2008.
52. S. Nandagopalan, C. Dhanalakshmi, B. Adiga, and N. Deepak (eds.), "A fast k-means algorithm for the segmentation of echocardiographic images using DBMS-SQL," in *2010 the 2nd International Conference on Computer and Automation Engineering (ICCAE)*, IEEE, 2010.
53. X. Li (ed.), "A volume segmentation algorithm for medical image based on k-means clustering," in *2008 IIHMSP'08 International Conference on Intelligent Information Hiding and Multimedia Signal Processing*, IEEE, 2008.
54. A. D. Kumbhar and A. Kulkarni (eds.), "Magnetic resonant image segmentation using trained k-means clustering," in *2011 World Congress on Information and Communication Technologies (WICT)*, IEEE, 2011.
55. K. Rathnayaka, T. Sahama, M. A. Schuetz, and B. Schmutz, "Effects of CT image segmentation methods on the accuracy of long bone 3D reconstructions," *Medical Engineering & Physics*, vol. 33, no. 2, 2011.
56. N. Sinha and A. Ramakrishnan (eds.) "Blood cell segmentation using EM algorithm," in *ICVGIP*, 2002.
57. H. Ng, S. Ong, K. Foong, P. Goh, and W. Nowinski (eds.), "Medical image segmentation using K-means clustering and improved watershed algorithm," in *2006 IEEE Southwest Symposium on Image Analysis and Interpretation*, IEEE, 2006.
58. M. Tabakov (ed.), "A fuzzy clustering technique for medical image segmentation," in *International Symposium on Evolving Fuzzy Systems*, IEEE, 2006.
59. D. N. Metaxas, *Physics-Based Deformable Models: Applications to Computer Vision, Graphics, and Medical Imaging*, Springer, 1996.
60. C. Xu, D. L. Pham, and J. L. Prince, "Image segmentation using deformable models," In *Handbook of Medical Imaging*, vol. 2, 2000, pp. 129–174.
61. R. Hegadi, A. Kop, and M. Hangarge, "A survey on the deformable model and its applications to medical imaging," *IJCA Special Issue on Recent Trends in Image Processing and Pattern Recognition*, pp. 64–75, 2010.
62. P. Sarkar, D. Chakrabarti, and M. Jordan, "Non-parametric link prediction," *arXiv* preprint arXiv: 11091077, 2011.
63. D. Heinz, *Hyper Markova Non-Parametric Processes for Mixture Modeling and Model Selection*, Carnegie Mellon University, 2010.
64. G. Tsechpenakis and D. Metaxas, "CoCRF deformable model: a geometric model driven by collaborative conditional random fields," *IEEE Transactions on Image Processing*, vol. 18, no. 10, pp. 2316–2329, 2009.
65. P. Yan and A. A. Kassim, "Medical image segmentation using minimal path deformable models with implicit shape priors," *IEEE Transactions on Information Technology in Biomedicine*, vol. 10, no. 4, pp. 677–684, 2006.

66. P. Yan and A. A. Kassim (eds.), "Medical image segmentation with minimal path deformable models," in *ICIP'04 2004 International Conference on IEEE Image Processing*, 2004.
67. M. Lee, S. Park, W. Cho, S. Kim, and C. Jeong, "Segmentation of medical images using a geometric deformable model and its visualization," *Canadian Journal of Electrical and Computer Engineering*, vol. 33, no. 1, pp. 15–19, 2008.
68. C. McIntosh and G. Hamarneh, "Medial-based deformable models in non-convex shape-spaces for medical image segmentation," *IEEE Transactions on Medical Imaging*, vol. 31, no. 1, pp. 33–50, 2012.
69. M. Farzinfar, E. K. Teoh, and Z. Xue (eds.), "A coupled implicit shape based deformable model for segmentation of MR images," in *2008 ICARCV 2008 10th International Conference on Control, Automation, Robotics, and Vision*, IEEE, 2008.
70. A. El-Baz and G. Gimel'farb (eds.), "Image segmentation with a parametric deformable model using the shape and appearance of priors," in *2008 CVPR 2008 IEEE Conference on Computer Vision and Pattern Recognition*, IEEE, 2008.
71. S. P. Awate, H. Zhang, and J. C. Gee, "A fuzzy, nonparametric segmentation framework for DTI and MRI analysis: with applications to DTItract extraction." *IEEE Transactions on Medical Imaging*, vol. 26, no. 11, pp. 1525–1536, 2007.
72. B. B. Nakhjavanlo, T. J. Ellis, P. Raoofi, J. Dehmeshki, "Medical image segmentation using deformable models and local fitting binary," *World Academy of Science*, vol. 52, pp. 168–71, 2011.
73. M. Sonka and J. M. Fitzpatrick (eds.), *Handbook of Medical Imaging* (Volume 2, Medical image processing and analysis), SPIE, The International Society for Optical Engineering, 2000.
74. N. Joshi and M. Brady, "Non-parametric mixture model based evolution of level sets and application to medical images," *International Journal of Computer Vision*, vol. 88, no. 1, pp. 52–68, 2010.
75. M. J. Dehghani, M. S. Helfroush, K. Kasiri, and K. Kazemi (eds.), "Atlas based segmentation of brain MR images using least square support vector machines," in *2nd International Conference on IPTA*, IEEE, 2010.
76. B. Van Ginneken, A. F. Frangi, J. J. Staal, B. M. ter Haar Romeny, and M. A. Viergever, "Active shape model segmentation with optimal features," *IEEE Transactions on Medical Imaging*, vol. 21, no. 8, pp. 924–933, 2002.
77. NUS S, "Active shape models," *Computer Vision and Image Understanding*, vol. 16, no. 1, 2012.
78. T. F. Cootes, G. J. Edwards, and C. J. Taylor, "Active appearance models," *IEEE Transactions on Pattern Analysis and Machine Intelligence*, vol. 23, no. 6, pp. 681–685, 2001.
79. P.-L. Chang and W.-G. Teng (eds.), "Exploiting the self-organizing map for medical image segmentation," in *2007 CBMS'07 Twentieth IEEE International Symposium on Computer-Based Medical Systems*, IEEE, 2007.
80. X. Chen, J. K. Udupa, U. Bagci, Y. Zhuge, and J. Yao, "Medical image segmentation by combining graph cuts and oriented active appearance models," *IEEE Transactions on Image Processing*, vol. 21, no. 4, pp. 2035–2046, 2012.
81. S. Zhan, H. Chang, J.-g. Jiang, and H. Li (eds.), "Spinal images segmentation based on improved active appearance models," in *2008 ICBBE the 2nd International Conference on Bioinformatics and Biomedical Engineering*, IEEE, 2008.

82. A. Neumann and C. Lorenz, "Statistical shape model based segmentation of medical images," *Computerized Medical Imaging and Graphics*, vol. 22, no. 2, pp. 133–143, 1998.
83. R. Smith and K. van Najarian (eds.), "Splines and active shape model for segmentation of pelvic x-ray images," in *2009 CME ICME International Conference on Complex Medical Engineering*, IEEE, 2009.
84. M. Rousson, N. Paragios, and R. Deriche, "Implicit active shape models for 3D segmentation in MR imaging," in *Medical Image Computing and Computer-Assisted Intervention–MICCAI 2004*, Springer, 2004, pp. 209–216.
85. C. Baillard, P. Hellier, and C. Barillot, "Segmentation of brain 3D MR images using level sets and dense registration," *Medical Image Analysis*, vol. 5, no. 3, pp. 185–194, 2001.
86. Y. T. Weldeselassie and G. Hamarneh (eds.), "DT-MRI segmentation using graph cuts," In *Medical Imaging*, International Society for Optics and Photonics, 2007.
87. J. Stawiaski, E. Decenciere, and F. Bidault (eds.), Interactive liver tumor segmentation using graph-cuts and watershed, in Workshop on 3D Segmentation in the Clinic: A Grand Challenge II Liver Tumor Segmentation Challenge MICCAI, New York, 2008.
88. B. Peng, L. Zhang, and D. Zhang, "A survey of graph theoretical approaches to image segmentation," *Pattern Recognition*, vol. 46, no. 3, pp. 1020–1038, 2013.
89. F. van der Lijn *et al.* (eds.), "Cerebellum segmentation in MRI using atlas registration and local multi-scale image descriptors," in *2009 ISBI'09 IEEE International Symposium on Biomedical Imaging: From Nano to Macro*, IEEE, 2009.
90. B. N. Li, C. K. Chui, S. Chang, and S. H. Ong, "Integrating spatial fuzzy clustering with level set methods for automated medical image segmentation," *Computers in Biology and Medicine*, vol. 41, no. 1, pp. 1–10, 2011.
91. N. Senthilkumaran and R. Rajesh, "Edge detection techniques for image segmentation—a survey of soft computing approaches," *International Journal of Recent Trends in Engineering*, vol. 1, no. 2, pp. 250–254, 2009.
92. S. Bhattacharyya, "A brief survey of color image preprocessing and segmentation techniques," *Journal of Pattern Recognition Research*, vol. 1, no. 1, pp. 120–129, 2011.
93. C. Jia-xin and L. Sen (eds.), "A medical image segmentation method based on watershed transform," in *2005 CIT 2005 The Fifth International Conference on Computer and Information Technology*, IEEE, 2005.
94. S. Li, L. Wu, and Y. Sun (eds.), "Cell image segmentation based on an improved watershed transformation," in *2010 International Conference on Computational Aspects of Social Networks (CASoN)*, IEEE, 2010.
95. R. Harini and C. Chandrasekar (eds.), "Image segmentation using nearest neighbor classifiers based on kernel formation for medical images," in *2012 International Conference on Pattern Recognition, Informatics and Medical Engineering (PRIME)*, IEEE, 2012.
96. M. Ravi and Ravindra S. Hegadi, "Detection of glomerulosclerosis in diabetic nephropathy using contour- based segmentation," *Procedia Science*, vol. 45, pp. 244–249, 2015.
97. M. Ravi and Ravindra S. Hegadi, "Detection of renal cell carcinoma—a kidney cancer using K-means clustering segmentation focused on pathological microscopic images," *Journal of Advanced Research in Dynamical and Control Systems*, pp. 144–149, 2017.

Index